Digital Database Architecture and Delineation Methodology for Deriving Drainage Basins, and a Comparison of Digitally and Non-Digitally Derived Numeric Drainage Areas

By Jean A. Dupree and Richard M. Crowfoot

Chapter 6 of
Section C, Geographic Information System Tools and Applications
Book 11, Collection and Delineation of Spatial Data

Technique and Methods 11–C6

U.S. Department of the Interior
U.S. Geological Survey

U.S. Department of the Interior
KEN SALAZAR, Secretary

U.S. Geological Survey
Marcia K. McNutt, Director

U.S. Geological Survey, Reston, Virginia: 2012

For more information on the USGS—the Federal source for science about the Earth, its natural and living resources, natural hazards, and the environment, visit http://www.usgs.gov or call 1–888–ASK–USGS.

For an overview of USGS information products, including maps, imagery, and publications, visit http://www.usgs.gov/pubprod

To order this and other USGS information products, visit http://store.usgs.gov

Suggested citation:
Dupree, J.A., and Crowfoot, R.M., 2012, Digital database architecture and delineation methodology for deriving drainage basins, and a comparison of digitally and non-digitally derived numeric drainage areas: U.S. Geological Survey Techniques and Methods, book 11, chap. C6, 59 p.

Contents

Figures

Tables

Conversion Factors

SI to Inch/Pound

Multiply	By	To obtain
	Length	
meter (m)	3.281	foot (ft)
kilometer (km)	0.6214	mile (mi)
	Area	
square meter (m^2)	10.76	square foot (ft^2)
square kilometer (km^2)	0.3861	square mile (mi^2)

Vertical coordinate information is referenced to the North American Vertical Datum of 1988 (NAVD 88).

Horizontal coordinate information is referenced to the North American Datum of 1983 (NAD 83).

Acronyms used in this report

CONUS	conterminous United States
CWSC	Colorado Water Science Center
DEM	digital elevation model
DRG	digital raster graphics
ESRI	Environmental Sciences Research Institute
FGDC	Federal Geographic Data Committee
GIS	geographic information system
GPS	global positioning system
HUC	hydrologic unit code
NAWQA	National Water Quality Assessment Program
NED	National Elevation Dataset
NHD	National Hydrography Dataset
NRCS	Natural Resources Conservation Service
NWIS	National Water Information System
SQL	Structured Query Language
USDA	U.S. Department of Agriculture
USGS	U.S. Geological Survey
UTM	Universal Transverse Mercator
WBD	National Watershed Boundary Dataset

Digital Database Architecture and Delineation Methodology for Deriving Drainage Basins, and a Comparison of Digitally and Non-Digitally Derived Numeric Drainage Areas

By Jean A. Dupree and Richard M. Crowfoot

Abstract

The drainage basin is a fundamental hydrologic entity used for studies of surface-water resources and during planning of water-related projects. Numeric drainage areas published by the U.S. Geological Survey water science centers in Annual Water Data Reports and on the National Water Information Systems (NWIS) Web site are still primarily derived from hard-copy sources and by manual delineation of polygonal basin areas on paper topographic map sheets. To expedite numeric drainage area determinations, the Colorado Water Science Center developed a digital database structure and a delineation methodology based on the hydrologic unit boundaries in the National Watershed Boundary Dataset. This report describes the digital database architecture and delineation methodology and also presents the results of a comparison of the numeric drainage areas derived using this digital methodology with those derived using traditional, non-digital methods.

Development of the digital database architecture included selecting the most appropriate data format, establishing the scope of the drainage basin archive, using a projection system that will yield the most accurate numeric drainage areas, and developing a suitable geodatabase architecture. A geographic information system (GIS) geodatabase structure was developed to archive drainage basin information as three GIS datasets: a point dataset to store site locations for which the basins are developed, a line dataset to retain information regarding the line source, and a polygon dataset to store the polygonal basin areas and corresponding numeric drainage area values. The basin delineation methodology discussed in this report involves the use of specific reference GIS datasets, specific delineation workflows, guidelines for interpreting topographic contours as drainage basins are delineated, and a protocol for finalizing each basin numeric drainage area for publication.

Numeric drainage areas derived by using the GIS method were compared to those derived using the manual delineation method, and any discrepancy between the two areal values was examined if the difference between the two was greater than 3 percent. A total of 518 drainage basins were delineated. Of 221 newly delineated basins with published NWIS numeric drainage areas, 21 had numeric drainage areas differing by more than 3 percent. Most discrepancies were found to be the result of the use of out-of-date paper maps, poor-quality maps, maps that are not contoured, or maps having a scale smaller than 1:24,000. Digitally derived polygonal basin areas were delineated in a fraction of the time compared to the traditional manual method and permit wider data access, and drainage basin comparisons can be based on an accessible, accurately known and agreed-upon basin entity.

Introduction

The drainage basin[1] is a fundamental hydrologic entity used for studies of surface-water resources and planning of water-related projects. Drainage basins are determined for various reasons, one of which is to provide basic basin properties (Office of Water Data Coordination, 1978). Two of these properties are foundational to many water studies. The first property, the polygonal basin area,[2] is the spatial extent of the drainage basin and is used to partition basin-specific characteristics (such as land-use, hydrologic, or climatic variables) for modeling, for statistical analysis of hydrologic quantities, and for analyses of the effects of various landscape variables on water movement and constituent transport. The second basin property, its numeric drainage area, is routinely derived by the U.S. Geological Survey (USGS) for streamgages and surface-water data-collection sites. The numeric drainage area is often used to normalize and compare basin-specific characteristics, such as population density or percent impervious surface, derived for multiple basins. The numeric drainage area is also used

---

[1]The term "drainage basin" will be used in this report to refer to a hydrologic region where potential surface runoff would converge, based on topography, to a single point in the landscape.

[2]Because the ambiguous term "area" by itself can refer to either a spatial entity or to a number, this report will use the phrase "polygonal basin area" to refer to the two-dimensional spatial extent of the drainage basin and "numeric drainage area" to denote the drainage-basin area value (in square miles, for example) computed for the spatial extent of a drainage basin.

for determining correlations and equations for investigations of water quality, geomorphology, or streamflow. Generally, numeric drainage area is also one of the most influential predictive variables in regression equations used to generate streamflow statistics, such as peak-flow or low-flow estimates (Capesius and Stephens, 2009). Such statistics are used to determine flood magnitude and extent, to estimate available water supply, to regulate contaminant loads, to formulate conservation strategies for aquatic species, for facilities planning, and to design transportation structures. Because polygonal basin areas and their derived numeric drainage areas are key elements, it is important that such entities be generated in a way that ensures their accuracy, consistency, and comparability.

At the time of this publication, hard-copy maps and notebook tabulations were still the most common source for USGS numeric drainage areas published in the USGS Annual Water Data Reports and on USGS NWISWeb (U.S. Geological Survey, 2002; U.S. Geological Survey, 2012a). During spring 2010, inquiries were made to USGS water science center offices to determine the origin of their published numeric drainage areas. Numeric-basin areas published in the Annual Water Data Reports and on NWISWeb for each state are most commonly (26 of 49 water science centers contacted) based on a combination of older polygonal basin areas determined using the hard-copy basin-delineation method and newer polygonal basin areas delineated using GIS techniques or StreamStats (Ries and others, 2008). Nine water science center offices based their published numeric drainage areas on polygonal basin areas derived using only hard-copy methods, and only 12 water science centers had a digital archive of most or all polygonal basin areas used to derive the numeric drainage areas published in the Annual Water Data Report or on NWISWeb. Two other offices stored all their polygonal basin areas in both a digital and a hard-copy archive. At one of these two offices, the hard-copy archive was still the foundation for numeric drainage areas published in the Annual Water Data Report and on NWISWeb and the GIS method was only used to check the published numeric drainage areas. Many USGS water science centers have an interest in transitioning from hard-copy storage of polygonal basin areas to digital archiving, and it is hoped that sharing a workable data structure and basin delineation methodology can serve a widespread USGS need.

Background

Historically, USGS water science centers have generated and recorded polygonal basin areas and numeric drainage areas using hard-copy materials, consisting of paper topographic maps and hand-written notebook entries. Using this methodology, referred to in this report as the "traditional method," polygonal basin areas are visually interpreted and hand traced on a paper topographic map, and a digital planimeter is used to determine numeric drainage

areas using published guidelines (Federal Interagency River Basin Committee, Subcommittee on Hydrology, 1951). For a drainage basin occupying several map sheets, a numeric drainage area is derived for the basin allotment on each sheet, and a total numeric drainage area for the basin is summed from aliquot areas on individual map sheets. The numeric drainage areas on individual map sheets, total numeric drainage areas, contributing numeric drainage areas, noncontributing numeric drainage areas, and revisions to numeric drainage areas are maintained as notebook entries, compiled by topographic quadrangle and tabulated by hydrologic accounting unit.

Shortcomings of the traditional method include the slow pace of work, the mixed quality and scale of topographic maps, the need for complex quality controls, and a relatively high potential for error. One shortcoming of this method is the variable map quality used to derive older numeric drainage areas. Maps used for tracing polygonal basin areas generally have a scale of 1:24,000; however, older smaller scale maps (1:100,000 or even 1:250,000) have been used in areas where 1:24,000-scale maps were not available at the time the polygonal basin area was first needed.

A particularly serious drawback to the traditional method is that hard-copy polygonal basin areas are effectively inaccessible, and because the polygonal basin areas generally span multiple topographic sheets in the topographic map archive, to view the entire area, several maps must be used. For most new drainage basins and most recent studies, USGS water-science-center personnel choose to generate a digital GIS-derived polygonal basin area, which can be created much more quickly but that may differ from the older, archived polygonal basin area derived by the traditional method. This situation leads to inconsistency between the newer GIS-derived polygonal basin areas and numeric drainage areas published in study reports and the analogous areas derived by the traditional method and published in the Annual Water Data Reports and on NWISWeb.

Two general GIS approaches are used to derive polygonal basin areas. One approach relies on the use of raster datasets to derive polygonal basin areas in an automated way. Several researchers have used automated methods to generate digital, polygonal basin areas using digital elevation model (DEM) data (Ellis and Price, 1995; Brown and others, 2000). A web-based example of the DEM-based method is USGS StreamStats (Ries and others, 2008). StreamStats is based on the use of pre-processed catchments derived from DEM[3] data that have been "conditioned" using two GIS datasets: (1) a stream network, which is "burned" as digital canyons in the elevation data, and (2) digital hydrologic unit boundaries, such as the National Watershed Boundary Dataset (WBD) (U.S. Geological Survey and U.S. Department of Agriculture, Natural Resources

---

[3]StreamStats uses either 10-m or 30-m DEM datasets, extracted from the National Elevaton Dataset (NED) (Gesch and others, 2002). The NED is derived from various data sources, often 1:24, 000-scale cartographic contours; see *http //ned.usgs.gov/Ned/faq.asp.*

Conservation Service, 2009), which are used to make digital "walls" of higher elevation cells that bound each delineated basin. As a StreamStats polygonal basin area is generated for a given location, catchments upstream from the location are rapidly combined, and the conditioned elevation data are used to trace the topographic drainage divide from the basin-site point to the nearest upstream catchment boundary. DEM-based basin delineation is quick, but can be inaccurate for small basins or in areas dominated by low relief.[4] Martucci and others (2006), for example, encountered problems with DEM-derived basins for coastal areas of the Chesapeake Bay drainage system, particularly for wider stream channels and for inaccurately delineated streams that were outside topographic lows. Also, all DEM-generated basin edges will be "jagged," reflecting the rectilinear edges of raster cells in the elevation data. Because the DEM available for each state can be 10 m or 30 m, and because states use different vector datasets (the WBD or a hybrid of WBD and in-house polygonal basin areas) to condition the DEM, it is difficult to specify a single delineation methodology based on StreamStats basins that will generate polygonal basin areas that conform to the Annual Water Data Report accuracy standards (Novak, 1985).

An alternative GIS approach is to vectorize polygonal basin areas by digitizing and interpreting drainage divides from topographic contours. If vector boundaries for hydrologic units such as the WBD are available, new basins can be delineated quickly by copying digital features and digitizing two new lines nearest the drainage basin-site point. This method is slower than the raster process, and ultimately based on human interpretation of topographic contour information (as opposed to the automated interpretation of contours used to generate DEM data), but polygonal basin areas generated using this approach precisely honor the vector boundaries generated from visual interpretation of the topographic contour information.

In 2003, the USGS National Water Quality Assessment Program (NAWQA) requested that the USGS Colorado Water Science Center (CWSC) delineate drainage basins for 85 streamgages in the South Platte River subregion. Because the traditional drainage basin delineation method is labor-intensive and time-consuming, a digital storage prototype (Dupree and Litke, 2005) for polygonal basin areas was developed for the CWSC. The USGS Virginia (Hayes and Weigand, 2007) and Maine (Pam Lombard, written commun., 2009) Water Science Centers later adopted the CWSC GIS architecture and GIS delineation methodology to derive drainage basins for stations. Polygonal basin areas for discharge-measurement locations and water-quality sampling

sites continued to be added to the CWSC prototype database until 2009, when the GIS archive was revised to be based on the WBD and expanded to include polygonal basin areas for all active Colorado streamgage locations.

Purpose and Scope

The purpose and scope of this report is to present a workable digital database design and basin delineation methodology that follows a GIS vectorization approach based on the WBD and that allows water science centers to store accurate numeric drainage areas for publication in the Annual Water Data Report and on NWISWeb. The report also presents a comparison of digitally derived numeric drainage areas with those derived using the traditional, manual method to determine their relative accuracy. Steps to create an appropriate digital storage structure and the GIS delineation workflow are summarized in this report and described in detail in Appendix 1, which serves as an instruction manual for creating a similar digital database structure and for populating it using the proposed WBD-based methodology. Guidelines for interpreting challenging topographic areas during basin delineation are provided in Appendix 2. Concurrently published with this report is a USGS Data Series (Dupree and Crowfoot, 2012) of the current USGS CWSC geodatabase as an example of the operational geodatabase template.

Methods

To test the accuracy and stability of the geodatabase data structure and delineation methodology, 518 drainage basins were delineated using 12-digit hydrologic unit boundaries from the WBD as a source. Such areas were delineated for sites located in all four major river basins in Colorado: the Arkansas River, the Colorado River, the Missouri River, and the Rio Grande basins (fig. 1). The database archive includes polygonal basin areas for 503 sites in Colorado, for 8 sites in Wyoming, for 6 sites in Nebraska, and for 1 site in Kansas. The digital archive includes polygonal basin areas for a variety of sites, including all active discharge measurement locations,[5] generally excluding sites located on manmade features such as canals or ditches. Numeric drainage areas in the digital archive range in size from 0.06 mi² to 24,943 mi².

Quality-control standards were incorporated into the digital data structure and delineation methodology. The design of the digital data architecture includes the use of a projection system that will provide accurate numeric drainage areas and also includes the implementation of spatial rules that constrain spatial interactions among features. As new polygonal basin areas are created, they are tested against these spatial rules, and if edits are needed they can be made to the shared boundary of multiple nested or adjacent basins to correct all basins simultaneously. Also, a protocol has been developed for basin review and finalization to ensure that published numeric drainage areas derived from the polygonal basin areas are as accurate as possible.

---

[4]To overcome this tendency locally in the areas of gages, some USGS water science centers occasionally add their own polygonal basin areas delineated from interpretation of topographic contours to the WBD as they prepare the "walling" dataset for StreamStats.

[5]Active as of September 30, 2009 and including streamgages where continuous measurements (real-time or data logger) are collected, crest-stage gages, discharge-measurement-only sites, and peak-flow-measurement sites at which there is a data collection platform.

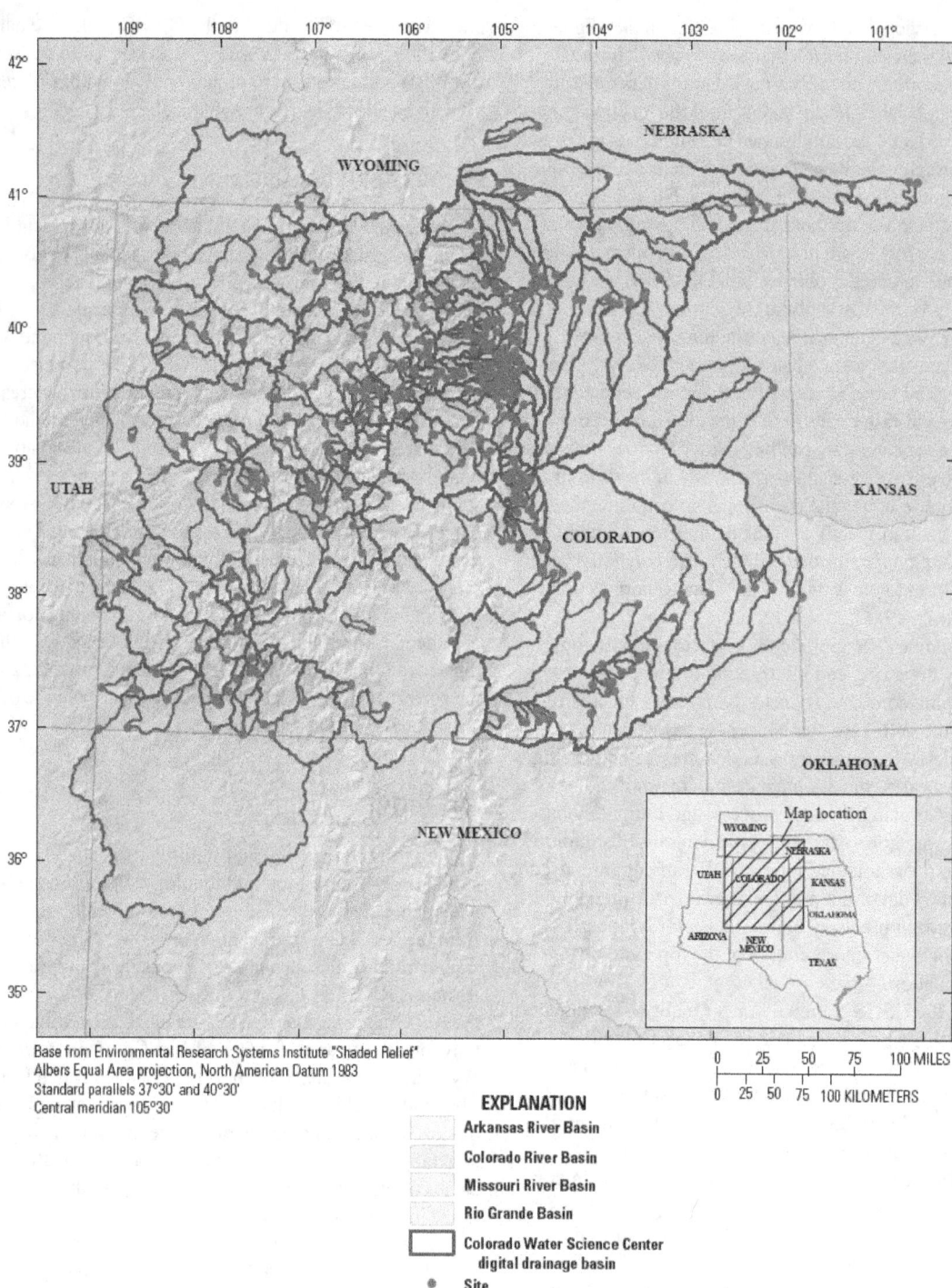

Figure 1.   Location map showing U.S. Geological Survey Colorado Water Science Center digital drainage basins.

Of the 518 basins delineated, 221 newly delineated basins had published USGS NWIS numeric drainage areas (derived by using hard-copy materials and the traditional method), and these published areas were compared to the numeric drainage areas for new digital polygonal basin areas. Where discrepancies exceeded 3 percent, the polygonal basin areas derived using both methodologies were inspected and probable reasons for the differences were determined.

Digital Database Architecture

This digital database architecture was developed by the USGS CWSC to determine numeric drainage areas to be published in the Annual Water Data Report and on the NWISWeb Web site. Development of the digital database architecture included establishing the data format and database scope for the drainage basin archive, using a projection system that will yield accurate numeric drainage areas, and developing a workable geodatabase schema.

Data Format and Database Scope

The data format chosen for archiving digital CWSC polygonal basin areas was the Environmental Sciences Research Institute (ESRI) file geodatabase. This GIS format was selected because file geodatabase size is limited only by available disk space and because it is optimized for storage and performance (Childs, 2009). An earlier ESRI geodatabase format (the personal geodatabase) is limited to a 2-gigabyte file size. File geodatabases are also easily displayed, edited, and manipulated in a desktop GIS environment. Additionally, because multiple USGS streamgages are frequently arrayed along a single stream, any GIS data format developed to store polygonal basin areas for such sites must be able to include nested drainage basins and permit editing of shared drainage-basin boundaries. The ESRI file geodatabase data structure allows inclusion and editing of topologically related, nested polygons.

The digital database developed by the USGS CWSC was constrained in terms of its geographic scope to include only polygonal basin areas that are wholly or partly within the State of Colorado. Occasional exceptions to this rule were made for basins needed for regional studies, such as those conducted

under the USGS NAWQA program. Polygonal basin areas stored in the CWSC archive also do not terminate at political boundaries but represent the full, hydrologic drainage basin.

The scope of the database was also limited to include only topographic (or geomorphic) drainage basins. Each topographic drainage basin is determined by the location of its site point, which should be located at or near the point of minimum elevation in the basin. The topographic drainage divide of the basin is the line of highest topographic elevations that directs potential surface runoff to the basin-site point. Topographic drainage basins are delineated using only elevation contours and without consideration of the water source. Such basins may have groundwater or diverted water leaving and entering across the drainage divide, but the polygonal basin area stored in the geodatabase is determined only by the surficial, topographic drainage divide. Polygonal basin areas in the CWSC geodatabase were delineated only for sites located near natural stream features and generally were not delineated for artificial features unless the feature represents a channelized stream that permanently diverts water flow. Topographic drainage basins may include areas of interior or noncontributing drainage. For the present CWSC geodatabase, only total polygonal basin areas were determined for hydrologic drainage basins, although contributing and noncontributing basin areas could be added as, or computed from, additional GIS datasets.

Geodatabase Projection System

The accuracy of numeric drainage areas depends on the use of a GIS projection system that does not distort area. An Albers equal-area-conic projection system was selected to ensure that all numeric drainage area measurements would be representative of actual areal measurements on the earth (Snyder, 1987). By using an equal-area projection system, the accuracy of derived numeric drainage areas becomes a function only of the accuracy of the delineated digital basin boundary.[6]

Although Universal Transverse Mercator (UTM) is a far more widely used projection system, a UTM projection will slightly distort areal and distance measurements.[7] For example, for a drainage basin located in central Colorado near the central meridian ($105°$) used for UTM Zone 13 north, the numeric drainage area in a UTM projection system differs from the numeric drainage area of the same drainage basin stored in an Albers equal-area-conic projection system by only 0.08 percent. However, differences between numeric drainage areas stored in the two projections increase with distance from the central meridian, with differences between numeric drainage areas as high as 0.18 percent in the eastern and western parts of Colorado. Raw differences between numeric drainage areas measured by these two projections also are larger for larger polygonal basin areas.

---

[6]Linear features bounding drainage basins are stored in the WBD and USGS CWSC geodatabase as both perimeter lines and as polygon boundaries. The term "basin boundary" is used in this report to refer collectively to both a drainage basin perimeter as well as the linear boundary of a basin polygon. The phrase "basin perimeter" will be used to refer only to basin-bounding features stored in the geodatabase as lines.

[7]Transverse Mercator projections, however, are "conformal" types of projection systems, meaning that they do not distort shapes (Snyder, 1987).

For example, the difference between the Albers equal-area-conic-projection-system numeric drainage area (19,009 mi²) and the UTM-Zone-13-projection numeric drainage area (18,997 mi²) for a large basin (South Platte River near Crook, CO) is about 12 mi². According to published guidance for the USGS Water Annual Data Report (Novak, 1985), published numeric drainage areas 1,000 mi² and greater are recorded to the nearest square mile.[8] Therefore, for such large basins, the UTM projection system may not be accurate enough for published areas in NWIS or in the Annual Water Data Report. Because the traditional method of basin delineation relies on paper topographic maps that have a UTM projection, this method potentially can generate inaccurate numeric drainage areas for larger basins. Although the Albers equal-area-conic projection system was selected for Colorado's geodatabase, UTM projections for geodatabases storing polygonal basin areas may be workable for other states if the basins are less than 1,000 mi² and are located near the central meridian of that UTM zone. Accurate areas for basins in such projections also can be calculated using available GIS tools.

Geodatabase Schema

A geodatabase schema defines the organizational structure of the geodatabase along with the rules, relationships, and properties of each GIS dataset stored in the geodatabase. The geodatabase stores GIS datasets, called feature classes, each of which includes features having a single type of geometry: point, line, or polygon.[9] The schema of the drainage basin geodatabase (Dupree and Litke, 2005) discussed in this report includes a single feature dataset that groups three feature classes: a point feature class that includes the site points for which each basin is developed, a line feature class that defines basin perimeters, and a polygon feature class that contains the polygonal basin areas. For certain fields in the feature classes, domains are used to restrict allowable field values, simplify data entry, and facilitate data queries and other analyses. Four persistent topology rules (Environmental Systems Research Institute, 2012a) were used and stored in the feature dataset: two that control the spatial quality of features in the line feature class and two additional rules that mandate that features in the line and polygon feature classes overlie one another.

Point Feature Class

The point feature class in this geodatabase contains the digital representation of the locations of the site points for which drainage basins are delineated. The basin-site point is the location of a streamgage orifice, a water-quality monitoring site, a stream confluence, or other USGS study- or project-related site location. Nearly all USGS NWIS site locations were derived from an NWIS sitefile location,

corresponding to the location shown on the NWISWeb site location map (U.S. Geological Survey, 2012a). For sites at stream confluences, digital topographic maps and current satellite imagery were used to locate the site point. Occasionally, however, the site point was not located precisely at the location shown on NWISWeb. Instead, it was located using other information, such as the site description given in the latest Annual Water Data Report that included the site, NWIS site coordinates, or consultation with field personnel familiar with the site location. NWIS sites that appeared to be incorrectly located were reviewed with USGS personnel responsible for editing NWIS locations in order to correct the NWIS site locations.

The point feature class has basic site information, including a site identifier ("SiteId") and a site name ("SiteName"). For USGS NWIS sites, the site identifier stored in the geodatabase is the USGS NWIS 8- to 15-digit site number; for confluences and other sites, it is an abbreviated identifier. The USGS 8-digit site number, when sorted, allows sites to be listed in downstream order for any major river basin; each 15-digit USGS NWIS site number is the concatenated latitude and longitude for the original site location, plus a 2-digit suffix. For confluences, the site identifier is a short name for the confluence. The site name stored in the geodatabase is either the formal USGS NWIS site name (which, for older streamgages, antedates the use of the USGS NWIS site number) or a descriptive name created for a confluence or other site. The geodatabase latitude "Latitude") and longitude ("Longitude") fields in the point feature class contain the decimal latitudes and decimal longitudes only for sites in the NWIS sitefile and are the coordinates in the NWIS sitefile; no latitude or longitude information is recorded in this feature class for non-NWIS sites. Other attributes that can be optionally added include the type of site ("SiteType"), the type of streamgage ("GageType"), and whether the site is an active streamgage ("Active"). Table 1 shows the attributes of the point feature class and examples of table entries.

Because the basin-site point, especially for streamgages, may be peripheral to the actual lowest basin elevation point for wide stream features, a second, optional, point feature class can be used to store the digital representation of the true topographic basin outlet, a feature that can be useful for hydrologic modeling applications. Points in the high-resolution (1:24,000-scale) National Hydrography Dataset (NHD) can be used to help generate locations for this optional feature class (U.S. Geological Survey, 2012b). These points can be annotated with NHD attributes that help link the point locations to the NHD.

---

[8]However, drainage areas larger than 1,000 square miles can be reported to less than three significant figures if the qualifying word "approximately" is used when publishing the numeric drainage area.

[9]Feature classes can also store specialized features such as annotation and dimension features.

Table 1. Attribute table for the point feature class showing examples of table entries.[1]

[Latitude in degrees north; longitude in degrees west of Greenwich Meridian]

OBJECTID	Shape	SiteId	SiteName	Latitude	Longitude	SiteType	GageType	Active
1	Point	09371000	MANCOS RIVER NEAR TOWAOC, CO	37.027	–108.741	Gage	Continuous-monitoring gage	Yes
274	Point	06718000	CLEAR CREEK BELOW 23RD AVENUE AT IDAHO SPRINGS, CO	39.742	–105.498	Gage and water quality	Continuous-monitoring gage	No
3	Point	07134100	BIG SANDY CREEK NEAR LAMAR, CO	38.114	–102.484	Gage	Continuous-monitoring gage	Yes
338	Point	394938104565300	SOUTH PLATTE RIVER AB CLEAR CR NR COMMERCE CTY, CO	37.324	–104.949	Water quality	Not applicable	No
216	Point	402114105350101	BIG THOMPSON BL MORAINE PARK NR ESTES PARK, CO	40.354	–105.584	Gage	Continuous-monitoring gage	Yes
430	Point	BuckhornMouth	BUCKHORN CREEK AT MOUTH	not used	not used	Confluence	Not applicable	No
7	Point	09144250	GUNNISON RIVER AT DELTA, CO	38.753	–108.078	Gage	Continuous-monitoring gage	Yes
72	Point	401723105400000	ANDREWS CREEK-LOCH VALE-RMNP	40.29	–105.667	Gage	Continuous-monitoring gage	Yes
310	Point	07113500	HUERFANO RIVER NEAR MUSTANG, CO	37.855	–104.701	Gage	Continuous-monitoring gage	No
477	Point	07126140	VAN BREMER ARROYO NEAR TYRONE, CO	37.399	–104.116	Gage	Crest-stage gage	Yes

[1]The Colorado Water Science Center geodatabase contains three additional, optional fields (shown) that indicate site type, gage type, and whether the site is active.

Line Feature Class

The line feature class contains basin perimeters. One attribute field was added to the line feature class to contain metadata about the line origin ("SourceOfLine"). This attribute for source of line contains two possible values that indicate whether the origin of the line was manual digitizing or the WBD. Metadata regarding the line origin is useful for several reasons. In general, only digitized lines are ever edited, and this attribute ensures that they can be distinguished from basin lines originating in the WBD. Also, annotating the lines in this way also allows for periodic comparisons of WBD-sourced lines against the WBD. Table 2 shows examples of entries in the table structure of this feature class and examples of table entries.

Table 2. Attribute table for the line feature class showing examples of table entries.

[Shape_length field in meters]

OBJECTID	Shape	SourceOfLine	Shape_Length
1	Polyline	National Watershed Boundary Dataset	43445.333
2	Polyline	National Watershed Boundary Dataset	14431.097
3	Polyline	National Watershed Boundary Dataset	5949.295
4	Polyline	Digitized line	2597.733
5	Polyline	National Watershed Boundary Dataset	21700.261
6	Polyline	Digitized line	740.975
7	Polyline	National Watershed Boundary Dataset	17946.063
8	Polyline	National Watershed Boundary Dataset	8942.268
9	Polyline	Digitized line	1095.269
10	Polyline	Digitized line	1405.917

Polygon Feature Class

The polygon feature class contains polygonal basin areas and records the derived numeric drainage areas. Each polygon in this feature class represents a separate basin. Most polygonal basin areas archived in the geodatabase are nested polygons for multiple gage locations arrayed along a single stream. The numeric drainage area for a basin polygon is automatically determined by the GIS software in the units specified by the projection as the "Shape_Area" field in the polygon feature-class table.

One feature of the geodatabase format that is different from the ESRI shapefile or ESRI coverage format is that it can store nested basins, and drainage basins developed for a series of USGS gages along a particular stream will be nested. Shapefiles also can store nested polygons as overlapping polygons, but these features are not related in any way topologically. The older ESRI coverage data format, in contrast, does store topological associations among features, but coverage polygons are not stored as nested features unless they are treated as a group of polygons, called a region.

Geodatabases store coincident, overlapping, and topologically related features, and therefore handle nested polygons well. Furthermore, the existing basic topological associations in geodatabase feature classes can be augmented by adding other topological rules.

The polygon feature class in this geodatabase includes several added attributes, including NWIS basin identifiers, fields that record the numeric drainage area, and information regarding the status of basin review. The NWIS site identifier ("SiteId") and NWIS site name ("SiteName") of the basin-site point are included in the polygon feature class so that it can be compared to the point feature class. Two fields are used to store numeric drainage areas. One is used to record the raw area in square miles ("AreaSqMi") calculated from the geodatabase polygon feature class "Shape_Area" field, and another area field stores the numeric drainage area rounded to NWIS standards ("AreaRounded"). A flag field is used during review of the polygonal basin area, and it is marked with a "Yes" if the polygonal basin area has been approved and the derived numeric drainage area is ready to be entered into NWIS. Optionally, another flag field can be added to record that the area actually has been entered or updated in NWIS ("NWISChanged"). Table 3 shows the table structure for the polygon feature class and examples of table entries.

Topology

ESRI geodatabases store coincident features. For example, a boundary common to two adjacent basin polygons is stored as two lines, not as a single line. If certain features require limits to constrain spatial interactions (if, for example, coincident lines need to overlie one another exactly), the geodatabase must incorporate spatial rules to control feature interrelations, called topology. Topology rules are spatial associations within or between feature classes that are used to find and fix spatial errors (MacDonald, 2001) and ensure that the spatial characteristics and interrelations of geodatabase features can be used to model real geographic features (in this case, drainage basins). By themselves, they do not prevent errors but are a screening mechanism used to examine the geodatabase to see if spatial discrepancies are present.

Geodatabase topology rules can be persistent or they can be created temporarily to check for errors and then deleted. The more topology rules, the more time is required to validate all the geographic features against the rules; therefore, some rules may be better utilized as periodic tests of the spatial quality of the data. For example, to eliminate short arcs that do not end at a line intersection, one can temporarily use the "Must Not Have Pseudonodes"[10] topology rule to find and merge short lines that do not have nodes on both ends (Hayes and Wiegand, 2007). Other topology rules, such as this one, may be added but the ones discussed in this report are the minimum ones needed.

---

[10]For clarity, ESRI's capitalization style for topology rules is followed in this report.

Table 3. Attribute table for the polygon feature class showing examples of table entries.

[Shape_Length field in meters; Shape_Area field in square meters]

Objectid	Shape	Siteid	SiteName	AreaSqMi[1]	AreaRounded[2]	Approved	NWISChanged	Shape_Length	Shape_Area
151	Polygon	09371000	MANCOS RIVER NEAR TOWAOC, CO	526.291	526.	Yes	Yes	235571.944	1363088589.145
221	Polygon	06718000	CLEAR CREEK BELOW 23RD AVENUE AT IDAHO SPRINGS, CO	258.088	258.	Yes	Yes	132348.767	668444011.581
62	Polygon	07134100	BIG SANDY CREEK NEAR LAMAR, CO	3215.559	3216.	Yes	Yes	629958.801	8328258112.936
48	Polygon	394938104565300	SOUTH PLATTE RIVER AB CLEAR CR NR COMMERCE CTY, CO	4085.577	4086.	Yes	Yes	581057.844	10581595910.426
364	Polygon	402114105350101	BIG THOMPSON BL MORAINE PARK NR ESTES PARK, CO	39.776	39.8	Yes	Yes	57891.503	103020336.719
264	Polygon	BuckhornCkMouth	BUCKHORN CREEK AT MOUTH	144.687	145.	Yes	Yes	107342.627	374736989.24
38	Polygon	09144250	GUNNISON RIVER AT DELTA, CO	5636.03	5636.	Yes	Yes	736457.403	14597250026.448
515	Polygon	401723105400000	ANDREWS CREEK-LOCH VALE-RMNP	0.655	0.66	Yes	Yes	5858.713	1696769.383
121	Polygon	07113500	HUERFANO RIVER NEAR MUSTANG, CO	791.332	791.	Yes	Yes	272385.767	2049539564.126
274	Polygon	07126140	VAN BREMER ARROYO NEAR TYRONE, CO	118.745	119.	Yes	Yes	88123.581	307549157.316

[1]Area is raw value, not rounded to USGS standards, converted from the Shape_Area field by dividing by 2,589,988.

[2]Area is rounded to USGS standards from AreaSqMi field.

Table 4 lists four persistent, topology rules that were used and stored in the feature dataset. Two rules limit spatial behavior of features in the line feature class: (1) a "Must Not Have Dangles" rule that screens drainage divides for any lines having dead ends, and (2) a "Must Not Intersect Or Touch Interior" rule, which ensures that each drainage divide will consist of a single line, and that all line intersections will have a point, or node, at the intersection. This latter rule can be helpful in detecting polygon slivers as new polygons are added to the basin archive. Two other topology rules determine whether the linear perimeters and polygon basin boundaries overlie one another and also allow them to be edited as a seamless piece. One rule demands that each line in the line feature class "Must Be Covered By Boundary Of" a polygon feature class,[11] and the second rule mandates, conversely, that each polygon feature's "Boundaries Must Be Covered By" a line in the line feature class. The latter rule combined with the line feature class "Must Not Intersect Or Touch Interior" rule (which prevents overlapping lines) unites multiple nested or adjacent polygon boundaries with a single corresponding line. A ramification of this rule is that, in order to invoke this rule to fuse adjacent and nested polygon boundaries during topological editing, it is helpful to establish a line feature class in addition to a polygon feature class to archive basin boundaries.

None of these rules alone prevent polygon overlaps, slivers, or gaps, which can be created if multiple, tightly spaced arcs occur along a basin boundary. The chance of encountering these problems is lessened by relying on a set basin delineation workflow—to add new basins one at a time and to use features from a single, consistent reference source, the WBD, when creating a new polygonal basin area.

A last, hidden (not listed in the topology rules) topology rule is called a cluster tolerance, which is set as the topology is created. Vertices separated by a distance less than the cluster tolerance are considered by the software to be spatially indistinguishable. After topology is validated, line vertices or point features within the cluster tolerance are shifted so that they have exactly the same coordinate values, and no two vertices will be closer than the cluster tolerance (MacDonald, 2001). The cluster tolerance of the basin geodatabase is set to match that of the WBD, which is used as a source for the locations of most topographic drainage divides stored in the basin archive.

Delineation Methodology

The delineation methodology and workflow developed for archiving USGS CWSC drainage basins is primarily based on the use of hydrologic unit boundaries available in a single, topologically consistent GIS dataset, the WBD. The delineation methodology is based on the use of reference GIS datasets, specific basin delineation workflows, guidelines for basin delineation, and a protocol used to finalize a numeric drainage area for publication.

Table 4.   Topology rules used in the basin geodatabase.

[N.A , not applicable]

Feature class	Topology rule	Feature class
Line feature class[1]	Must Not Have Dangles[2]	N.A.
Line feature class	Must Not Intersect Or Touch Interior	N.A.
Line feature class	Must Be Covered By Boundary Of	Polygon feature class
Polygon feature class	Boundary Must Be Covered By	Line feature class

[1]The topology rule is read as a sentence from left to right: "Line feature class Must Not Have Dangles."

[2]The topology rules follow the capitalization used by the Environmental Systems Research Institute (ESRI).

Reference GIS Datasets

Two reference GIS datasets were used for basin delineation: the Natural Resources Conservation Sevice (NRCS) WBD and the USGS 1:24,000-scale digital raster graphic (DRG) topographic maps. The primary reference GIS dataset for basin delineation was the WBD, a National Spatial Data Infrastructure Data Theme of 12-digit hydrologic unit boundaries developed using the USGS 1:24,000-scale DRG maps as the ultimate reference for line work. The WBD was developed under the leadership of the Federal Geographic Data Committee (FGDC) Subcommittee on Spatial Water Data to conform to National Map Accuracy Standards (U.S. Geological Survey and U.S. Department of Agriculture, Natural Resources Conservation Service, 2009). It is a nationally consistent, seamless geospatial database. The 12-digit hydrologic unit boundaries in the WBD have been certified and finalized for all states in the conterminous United States and are available both as national and state-wide GIS datasets. Basins delineated in the CWSC geodatabase were derived as much as possible by copying existing lines from this single digital data source. In situations where errors are found in the WBD, these can be submitted to the WBD stewards for updates to this GIS dataset.[12]

In compliance with FGDC standards for delineating hydrologic unit boundaries (Federal Geographic Data Committee, 2003), the USGS 1:24,000-scale DRG topographic maps are the base reference for all polygonal basin areas in the geodatabase, including those derived from the WBD. Most polygonal basin areas in the geodatabase described in this report consist of copied WBD features; however, 1:24,000-scale DRG maps were used to digitize new lines, and generally two digitized lines connect the basin-site point to the nearest upstream WBD feature.

[11]Each topology rule is read properly as a sentence. The feature classes are "nouns" and the topology rule is a "verb" that defines what the feature classes are to "do" in a spatial sense.

[12]Names of the WBD stewards for each state are available at the NHD Web site (U.S. Geological Survey, 2012c).

Basin Delineation Workflows

Using the WBD and the DRG maps, drainage basins were delineated for 518 Colorado sites, including 291 active Colorado streamgages that are not located on diversion structures. Polygonal basin areas were added sequentially to the geodatabase to allow the delineator to catch any topological errors that may have been produced during basin delineation and also to help prevent slivers and gaps between the new basin polygon and those already stored in the permanent digital archive.

Two workflows (both described in Appendix 1) were developed for basin delineation: one a method by which a new polygonal basin area is made by creating a basin perimeter from the WBD and digitized lines and the other in which WBD polygons are combined and cut to a final basin shape. To generate a polygonal basin area using the first method, WBD lines and digitized lines are assembled and trimmed until they form a single closed loop, and this perimeter loop is then used to construct a new basin polygon. For the second method, WBD basin polygons upstream from, and including, the basin site location are merged, a cutter line is created and used to cut the merged polygon into two pieces, and the upstream-most piece is retained as the new basin polygon.

Basin Delineation Guidelines

Two key controls for determining an accurate numeric drainage area are a correct site-point location and the accurate positioning of the topographic drainage divide. Federal and USGS standards for basin delineation, especially for topographic interpretation (Federal Interagency River Basin Committee, Subcommittee on Hydrology, 1951; Federal Geographic Data Committee, 2003; U.S. Geological Survey and U.S. Department of Agriculture, Natural Resources Conservation Service, 2009), are utilized and other delineation guidelines were developed for the CWSC geodatabase (Appendix 2).

Prior to delineating any basin, the location of the basin-site point (whether a streamgage, water-quality sampling site, confluence, or other location) is verified. If the site is a USGS NWIS site, it is compared to the NWISWeb location map or from the published site description in the USGS Annual Water Data Report. In several cases, the actual NWIS site location was verified by consulting hydrologic field personnel. If the polygonal basin area is being delineated for a confluence or another type of site, the site-point location is generally referenced to current satellite imagery.

Topographic drainage divides are delineated to be accurate at a map scale of 1:24,000 using the two reference GIS datasets: (1) lines and polygons in the WBD and (2) DRG topographic map contours. New lines are digitized from DRG maps by starting at the site point on the stream for which the polygonal basin area is being derived and following the topographic drainage divides indicated by elevation contours (Federal Interagency River Basin Committee, Subcommittee on Hydrology, 1951). In-house accuracy standards have been developed to make the polygonal basin areas more reliable for water resource analyses. Where the topographic drainage divide between two drainage basins is sharp, an attempt is made to position the line within 1/8th inch of the ridgeline, at a scale of 1:10,000 or greater. Such accuracy is needed because digitized lines can potentially bound two adjacent basins and several nested basins.

Several additional in-house delineation guidelines for topographic interpretation of problematic areas, given in Appendix 2, were developed from reference publications that address standards to be used to delineate USGS drainage basins (Federal Interagency River Basin Committee, Subcommittee on Hydrology, 1951; Office of Water Data Coordination, 1978; Federal Geographic Data Committee, 2003; U.S. Geological Survey and U.S. Department of Agriculture, Natural Resources Conservation Service, 2009). Most such in-house guidelines concern topographic interpretation, such as how to draw drainage divides near anthropogenic features (such as railroad and highway berms) and non-anthropogenic features (such as dune fields).

Finalizing the Delineated Basin

One of the most important quality controls is the review process used to confirm the accuracy of each polygonal basin area and the derived numeric drainage areas, which is performed by comparison to a digital stream network and then to published numeric drainage areas. After the drainage basin was delineated, it was compared to a digital stream network, the medium- or high-resolution NHD. The polygonal basin area was checked to ensure that it surrounded the digital stream network draining to the basin site-point location.

Numeric drainage areas developed for sites that already have published numeric drainage areas in NWIS are compared to the NWIS numeric drainage area. Generally both a raw difference and a percentage difference threshold are used for such comparisons. Basins that have a raw difference in the arbitrary range of 10 mi^2 to 30 mi^2 generally are visually compared to topographic map sheets and notebooks. Those having a percent difference in excess of 3 percent are also inspected to determine a reason for the area discrepancy.

As a final check, the numeric drainage area of the new drainage basin is compared with available, published NWIS numeric drainage areas for sites upstream and downstream. Newly digitized line features are also checked to verify the topographic interpretation. After this comparison, if the new numeric drainage area seems reasonable, the new basin is marked in the polygon feature class table as approved for publication on NWISWeb or in the Annual Water Data Report.

Comparison of Digitally and Non-Digitally Derived Numeric Drainage Areas

GIS-derived numeric drainage areas were compared to those derived using the traditional method, and if the difference between the GIS-derived drainage area and the published NWIS drainage area (U.S. Geological Survey, 2012a) exceeded 3 percent, the two boundaries were compared to determine a reason for the discrepancy. A total of 518 drainage basins were delineated using the GIS methodology, and 221 of the newly delineated drainage basins had published NWIS numeric drainage area. Absolute values of the percent differences between the NWIS and GIS numeric drainage areas were between 0.008 and 35 percent, with a median value of 0.29 percent. More than 78 percent of the 221 newly delineated polygonal basin areas had less than 1 percent difference between their derived numeric drainage areas. In a cumulative frequency plot of the distribution of the absolute value of the percent difference, a data breakpoint was found to occur around 3 percent (fig. 2). Of 221 basins that had published NWIS numeric drainage areas, the areas of 21 were found to differ from published NWIS numeric drainage area by more than this 3 percent threshold (table 5). Several tabulated discrepancies occurred in groups along a main stem due to the propagation of an error from an upstream basin to those basins to which it is tributary.

Of the 21 numeric drainage-area discrepancies,[13] 10 (error code "M" in table 5) were wholly or partly caused by the use of out-of-date paper maps, poor-quality maps, maps that are not contoured, or maps having a smaller scale

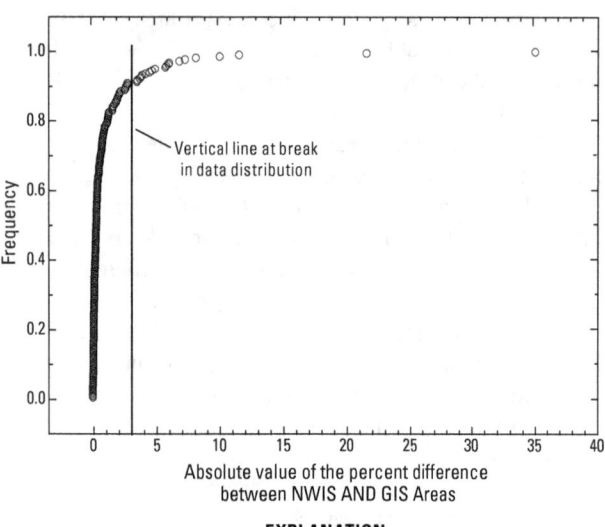

Figure 2. Cumulative frequency plot of the absolute value of the percent difference between non-digital and digital areas.

than 1:24,000. In several instances, an old numeric drainage area determined from small-scale, older topographic maps was inherited by a downstream site, and the older numeric drainage area was appended to a newer numeric drainage area determined for the intervening polygonal basin area delineated from 1:24,000-scale maps. One-half of such map-caused errors were found for large basins with areas greater than 1,000 mi^2, where it would have been time-consuming to re-delineate a polygonal basin area based on newer, 1:24,000-scale paper maps. For several basins, the error also involved the adoption or incorporation of older numeric drainage areas published by other federal agencies[14] that, themselves, relied on old, small-scale maps for basin delineation.

In seven instances, discrepancies were caused by errors in the hard-copy recording process. In one case, for the "Purgatoire River at Trinidad, CO" site, updates to the numeric drainage area did not propagate to the NWIS sitefile (error code "S"). Tabulation errors (errors by which the aliquot numeric drainage areas on topographic map sheets are tabulated in notebooks) occurred during the delineation of a group of five drainage basins along the Purgatoire River in southeastern Colorado, because a section of a basin had been delineated on the adjacent New Mexico topographic maps but omitted from notebook area totals (error code "T").

Seven numeric drainage areas were generated differently by the two respective methods because noncontributing polygonal basin areas were allocated differently (error code "N"). Several downstream streamgages in the Purgatoire River drainage basin (including Purgatoire River and Van Bremer Arroyo sites) had a noncontributing polygonal basin area that was adjacent to the basin but not a part of the Purgatoire hydrologic unit nor a part of the hydrologic unit to the west. Polygonal basin areas delineated using the traditional method for the Purgatoire River included that adjoining noncontributing basin area, which, in the WBD, was part of the adjacent hydrologic unit to the west.

Other sporadic types of delineation-related errors were found. For the site at "Joe Wright Creek above Joe Wright Reservoir, CO," a small portion of the polygonal basin area recorded on a 1:24,000-scale topographic map was incorrectly delineated (error code "D"). The polygonal basin area for "South Platte River above Cheeseman Lake, CO," was incorrectly delineated using the GIS method and later corrected (error code "G"). Both the hard-copy- and GIS-derived polygonal basin area were also initially incorrect for the basin delineated for "Sand Creek at Colorado-Wyoming State line", because the basin outlet site was incorrectly located (error code "O").

---

[13]Several numeric drainage-area discrepancies had multiple causes; therefore, the totals cited in these paragraphs will not sum to the total number of basins (21) listed in table 5.

[14]Such as numeric drainage areas published by the U.S. Army Corps of Engineers, derived from topographic maps in use before the early 1950s.

Table 5. Comparison of geographic information systems (GIS) and published National Water Information System (NWIS) numeric drainage areas that differed by more than 3 percent.

[D, delineation error - traditional method; G, GIS delineation error - GIS method; M, small-scale map or poor map quality - traditional method; N, different allocation of non-contributing area - both methods; O, outlet location error - both methods; S, sitefile not updated - traditional method; T, tabulation error, portion of basin not included in notebook tabulations - traditional method]

SiteId	SiteName	[1]NWIS area, mi²	GIS area, mi²	[2]Difference, mi²	[3]Percent difference	Error code
07119500	APISHAPA RIVER NEAR FOWLER, CO	1,125.	1,070.	55.	4.9	M
07094500	ARKANSAS RIVER AT PARKDALE, CO	2,548.	2,459.	89.	3.5	M
06746095	JOE WRIGHT CREEK ABOVE JOE WRIGHT RESERVOIR, CO	3.01	3.14	−0.13	4.2	D
09260000	LITTLE SNAKE RIVER NEAR LILY, CO	3,730.	4,034.	−304.	8.2	M
09253000	LITTLE SNAKE RIVER NEAR SLATER, CO	285.	252.	33.	11.6	M
09349800	PIEDRA RIVER NEAR ARBOLES, CO	629.	653.	−24.	3.8	M
07128000	PURGATOIRE R AT HIGHLAND DAM, NR LAS ANIMAS, CO	3,203.	3,328.	−125.	3.9	T, N
07126500	PURGATOIRE RIVER AT NINEMILE DAM, NR HIGBEE, CO	2,752.	2,875.	−123.	4.5	T, N
07126485	PURGATOIRE RIVER AT ROCK CROSSING NR TIMPAS, CO	2,635.	2,758.	−123.	4.7	T, N
07124500	PURGATOIRE RIVER AT TRINIDAD, CO	795.	749.	46.	5.8	S
07128500	PURGATOIRE RIVER NEAR LAS ANIMAS, CO	3,318.	3,441.	−123.	3.7	T, N
07126300	PURGATOIRE RIVER NEAR THATCHER, CO	1,791.	1,914.	−123.	6.9	T, N
09342500	SAN JUAN RIVER AT PAGOSA SPRINGS, CO	298.	281.	17.	5.7	M
06659580	SAND CREEK AT COLORADO-WYOMING STATE LINE	29.2	30.8	−1.6	5.5	O
09255000	SLATER FORK NEAR SLATER, CO	161.	151.	10.	6.2	M
06700000	SOUTH PLATTE RIVER ABOVE CHEESMAN LAKE, CO	1,628.	1,720.	−92.	5.6	G
07126200	VAN BREMER ARROYO NEAR MODEL , CO	175.	162.	13.	7.4	N
07126140	VAN BREMER ARROYO NEAR TYRONE, CO	132.	119.	13.	9.8	N
07134990	WILD HORSE CREEK ABOVE HOLLY, CO	270.	175.	95.	35.2	M
09260050	YAMPA RIVER AT DEERLODGE PARK, CO	7,660.	7,931.	−271.	3.5	M
09247600	YAMPA RIVER BELOW CRAIG, CO	1,750.	2,128.	−378.	21.6	M

[1]Formerly published drainage area; published NWIS drainage areas have been updated.

[2]NWIS area - GIS area.

[3]Absolute value((NWIS area - GIS area) / NWIS area) * 100).

In general, the care and attention to detail by individuals who generated polygonal basin areas using the traditional method was good. The accuracy of manually derived CWSC polygonal basin areas was good considering the quality of topographic or other maps generally available during the era when many of the larger polygonal basin areas were generated. It should be noted, again, that because the projection system used for USGS topographic maps is UTM, polygonal basin areas for larger basins delineated even on good-quality topographic maps will not be as accurately determined as they would be using an equal-area projection system. Creation of a digital archive does not preclude the continued use of the hard-copy archive; in fact, it was found that for some basins in urban areas, having a historical version of the topography that pre-dated urbanization often helped identify where the polygonal basin area should be delineated because topographic contours can be harder to discern on more recent topographic maps of urban areas.

The GIS methodology of deriving polygonal basin areas and numeric drainage areas, however, offers the advantages of a uniform digital map-scale and speed in delineation and computation. Polygonal basin areas derived using GIS are a seamless, unified piece, as opposed to hardcopy storage in which the polygonal basin area is fragmented because it is stored on multiple map sheets. Drainage divides can be traced accurately on screen using DRG topographic maps, which have a common scale and datum, unlike the paper archive maps, which can vary in scale, datum, source, and quality. Numeric drainage areas are calculated automatically by the GIS software for polygon feature classes and any refinements to a boundary trigger an automatic recalculation of the numeric drainage area. A polygonal basin area can be delineated quickly (generally, within a half-hour) compared to the delineation of only a dozen basins per year using the traditional method. Digital polygonal basin areas also permit wider data access, and basin comparisons can be based on an accessible, accurately known, and agreed-upon basin entity.

Summary

The drainage basin is a fundamental entity used for studies of surface-water resources and during planning of water-related projects. Historically, the U.S. Geological Survey (USGS) water science centers have recorded polygonal basin areas and numeric drainage areas using hard-copy materials, consisting of paper topographic maps and hand-written notebook entries. This report presents a workable digital database design and delineation methodology to store published USGS polygonal basin areas that are derived in a consistent way from a uniform source. This digital database architecture was developed by the USGS Colorado Water Science Center (CWSC) to determine numeric drainage areas published in the USGS Annual Water Data Report and on the USGS National Water Information System (NWIS) Web site. The report also presents the results of a comparison of digitally derived numeric drainage areas with those derived using the traditional, manual method to determine the relative accuracy of both methods.

Development of the digital database architecture included establishing the most appropriate data format and scope of the basin archive, using a projection system that yielded accurate numeric drainage areas, and developing a workable geodatabase schema. The Environmental Research Systems Institute (ESRI) file geodatabase was chosen as the geographic information system (GIS) data format to archive the digital CWSC polygonal basin areas. Because the accuracy of basin numeric drainage areas depends on the use of a projection system that minimizes distortion, an Albers equal-area-conic projection system was used. The schema used for the drainage basin geodatabase includes a single feature dataset that groups three GIS feature classes: a point feature class that includes the site locations for which each drainage basin is developed, a line feature class that defines basin perimeters, and a polygon feature class that contains the polygonal basin areas.

The basin delineation methodology uses reference digital GIS datasets, specific delineation workflows, guidelines for basin delineation, and a protocol for finalizing each numeric drainage area for publication. Two reference GIS datasets were used for basin delineation: the Natural Resources Conservation Service National Watershed Boundary Dataset (WBD) and the USGS 1:24,000-scale digital raster graphics (DRG) topographic maps. Using the WBD and the DRG maps, drainage basins were delineated for 518 Colorado sites, including active Colorado streamgages that are not located on diversion structures. Two workflows were developed for basin delineation: one a method which derives a basin perimeter from the WBD and digitized lines, and the other a method which cuts a basin polygon out of merged WBD polygons. Several additional in-house delineation guidelines for topographic interpretation in problematic areas were developed. Two key controls for determining an accurate numeric drainage area are a correct basin site location and accurate positioning of the drainage divides. One of the most important quality controls is the final review process used to confirm the accuracy of each polygonal basin area, which is performed by comparison to a digital stream network followed by a comparison to published numeric drainage areas.

GIS-derived numeric drainage areas were compared to those derived using the traditional method, and any discrepancy between the numeric areas was examined if the difference exceeded a threshold of 3 percent. Of the 518 basins delineated, 221 had published numeric drainage areas in NWIS. Of these 221 basins, 21 had numeric drainage areas differing by more than this threshold. Most such discrepancies were found to be wholly or partly caused by the use of out-of-date paper maps, poor-quality maps, maps that are not contoured, or maps having a smaller scale than 1:24,000. Digitally derived basins were delineated in a fraction of the time compared to the traditional method and permit wider data access, and basin comparisons can be based on an accessible, accurately known, and agreed-upon basin entity.

Acknowledgments

The development of this digital-basin-delineation methodology was done by the Colorado Water Science Center in cooperation with the USGS NAWQA program and the USGS Central Region Director's Office.

The authors thank the following people: Jennifer Krstolic and Donald Hayes, of the USGS Virginia Water Science Center, both contributed very useful insights and suggestions for improvement for the delineation methodology. Kristine Verdin, GIS Coordinator for the USGS CWSC, contributed useful information and feedback during the report review. Luther Schalk and Pam Lombard of the USGS Maine Water Science Center also provided valuable feedback regarding the early data structure of the CWSC geodatabase. The authors contacted the USGS water science center for each state to determine how they derived published numeric drainage areas. The individuals who responded were very generous with their time and comments that were invaluable to this report. Curtis Price of the South Dakota Water Science Center was helpful and informative in discussing the relative accuracy of equal-area projections compared to other projection systems.

References Cited

Brown, J.R., Ulery, R.L., Parcher, J.W., 2000, Creating a standardized watershed database for the Lower Rio Grande/Rio Bravo, Texas: U.S. Geological Survey Open-File Report 00–065, 17 p.

Capesius, J.P., and Stephens, V.C., 2009, Regional regression equations for estimation of natural streamflow statistics in Colorado: U.S. Geological Survey Scientific Investigations Report 2009–5136, 46 p. (also available at: *http://pubs.usgs.gov/sir/2009/5136/*).

Childs, C., 2009, Top nine reasons to use a file geodatabase—A scalable and speedy choice for single users and small groups: Environmental Systems Research Institute white paper, accessed January 7, 2012, at *http://www.esri.com/news/arcuser/0309/files/9reasons.pdf*.

Dupree, J.A., and Crowfoot, R.M., 2012, Geodatabase of digital basin boundaries developed by the U.S. Geological Survey Colorado Water Science Center, published 2012: U.S. Geological Survey Data Series 650.

Dupree, J.A., and Litke, D.W., 2005, Drainage area determination using ArcGIS8 and the geodatabase, *in* Char, S.J., and Sieverling, J.B., eds., 2005, Proceedings of the U.S. Geological Survey Fifth Biennial Geographic Information Systems Science Workshop, March 1–5, 2004, Denver, Colorado: U.S. Geological Survey Scientific Investigations Report 2005–5236, p 14.

Ellis, W.H., and Price, C.V., 1995, Development of a 14-digit hydrologic coding scheme and boundary data set for New Jersey: U.S. Geological Survey Water-Resources Investigations Report 95–4134, 1 sheet.

Environmental Systems Research Institute, 2012a, ArcGIS Geodatabase Topology Rules, accessed January, 7, 2012, at *http://webhelp.esri.com/arcgisdesktop/9.3/pdf/Topology_rules_poster.pdf*.

Environmental Systems Research Institute, 2012b, ESRI Maps and Data, accessed January, 28, 2012 at *http://www.arcgis.com/home/group.html?q=tags:ArcMap931_Base&t=group&owner=esri&title=ESRI%20Maps%20and%20Data&sortField=title&sortOrder=asc&content=all*.

Federal Geographic Data Committee, 2003, Federal Standards for Delineation of Hydrologic Unit Boundaries: Federal Geographic Data Committee Proposal, Version 1.1, 53 p.

Federal Interagency River Basin Committee, Subcommittee on Hydrology, 1951, Interagency coordination of drainage area data: Notes on hydrologic activities, Bulletin No. 4, 49 p.

Gesch, D., Oimoen, M., Greenlee, S., Nelson, C., Steuck, M., and Tyler, D., 2002, The national elevation dataset: Photogrammetric Engineering and Remote Sensing, v. 68, no. 1, p. 5–11.

Hayes, D.C. and Wiegand, U., 2007, Drainage areas of selected streams in Virginia: U.S. Geological Survey Open-File Report 2006–1308, 51 p.

Horizon Systems Corporation, 2012, NHDPlus Home, accessed January 7, 2012, at *http://www.horizon-systems.com/nhdplus/*.

MacDonald, A., 2001, Building a Geodatabase: Environmental Systems Research Institute, Inc., Redlands, California, 481 p.

Martucci, S.K., Krstolic, J.L., Raffensperger, J.P., and Hopkins, K.J., 2006, Development of land segmentation, stream-reach network, and watersheds in support of Hydrological Simulation Program-Fortran (HSPF) modeling, Chesapeake Bay Watershed, and Adjacent Parts of Maryland, Delaware, and Virginia: U.S. Geological Survey Scientific Investigations Report 2005–5073, 19 p.

Novak, C.E., 1985, WRD Data reports preparation guide: U.S. Geological Survey, 199 p.

Office of Water Data Coordination, 1978, Physical basin characteristics for hydrologic analysis, *in* National handbook of recommended methods for water-data acquisition: Reston, Va., U.S. Geological Survey, Chapter 7, 38 p.

Ries, K.G., III, Guthrie, J.G., Rea, A.H., Steeves, P.A., Stewart, D.W., 2008, StreamStats: a water-resource web application: U.S. Geological Survey Fact Sheet 2008–3067, 6 p.

Snyder, J.P., 1987, Map projections—A working manual: U.S. Geological Survey Professional Paper 1395, 383 p.

U.S. Department of Agriculture Natural Resources Conservation Service, 2012a, GeoSpatial Data Gateway: accessed January 7, 2012, at *http://datagateway.nrcs.usda.gov/*.

U.S. Department of Agriculture Natural Resources Conservation Service, 2012b, FTP directory: accessed January 7, 2012, at *ftp://ftp.ftw.nrcs.usda.gov/pub/wbd/*.

U.S. Geological Survey, 2002, NWISWeb: new site for the Nation's water data: U.S. Geological Survey Fact Sheet 128–02, 2p.

U.S. Geological Survey, 2012a, USGS National Water Information Systems Web Interface (NWISWeb): accessed January 7, 2012, at *http://waterdata.usgs.gov/nwis*.

U.S. Geological Survey, 2012b, National Hydrography Dataset: accessed January 7, 2012, at *http://nhd.usgs.gov/data.html*.

U.S. Geological Survey, 2012c, Watershed Boundary Dataset (WBD) State Stewards: accessed January 28, 2012, at *http://nhd.usgs.gov/wrd_stewardship.html*.

U.S. Geological Survey, 2012d, USGS Publications Warehouse: accessed January 7, 2012, at *http://pubs.er.usgs.gov/*.

U.S. Geological Survey, 2012e, Welcome to StreamStats: accessed January 9, 2012, at *http://water.usgs.gov/osw/streamstats*.

U.S. Geological Survey, 2012f, The National Map: accessed January 9, 2012, at *http://nationalmap.gov*.

U.S. Geological Survey and U.S. Department of Agriculture, Natural Resources Conservation Service, 2009, Federal guidelines, requirements, and procedures for the National Watershed Boundary Dataset *in* chap. 3 of Section A, Federal Standards, Book 11, Collection and Delineation of Spatial Data: 68 p. (Also available at *ftp://ftp-fc.sc.egov.usda.gov/NCGC/products/watershed/hu-standards.pdf*).

Wiki.GIS.com, 2012, Albers equal-area conic projection: accessed January 7, 2012, at *http://wiki.gis.com/wiki/index.php/Albers_equal-area_conic_projection*.

XToolsPro, 2012, XToolsPro: accessed January 30, 2012, at *http://www.xtoolspro.com/*.

Appendixes

Appendix 1.   Procedures and Guidelines for Delineating Basins

Introduction

This appendix includes five sections. The first section presents steps for creating a geodatabase and geodatabase components used to store polygonal basin areas and their derived numeric drainage areas. The second section describes the digital datasets that are useful during drainage basin delineation. A third section discusses locating the site point, which is crucial to subsequent basin delineation. The fourth section describes basin delineation and provides two workflows, general delineation guidelines, and procedures for finalizing polygonal basin areas and numeric drainage areas. The last section provides a discussion of common editing problems and solutions, particularly those that can arise after basins have been added to the geodatabase archive. A U.S. Geological Survey (USGS) Data Series of the Colorado Water Science Center (CWSC) geodatabase is available for reference (Dupree and Crowfoot, 2012).

Creating the Basin Geodatabase and Its Components

This first section presents instructions for creating a geodatabase structure and geodatabase components that can be used to archive polygonal basin areas. Included are the steps to create the file geodatabase; the attribute domains; the basin feature dataset; the point, line, and polygon feature classes; and the appropriate topology rules. It is assumed that the reader has a basic familiarity with geographic information systems (GIS) and the ArcGIS environment.

Creating the Geodatabase

A geodatabase is a relational database that stores geographic data. Its basic function is to group and organize spatial data tables, non-spatial data tables, and rule tables. File geodatabases can be created in the Environmental Sciences Research Institute (ESRI) ArcCatalog[15] interface using either the context menu (accessible by right clicking on any existing folder name in the ArcCatalog tree) or the ArcCatalog file drop-down menu. For either method, the user selects "New" and then "File Geodatabase." Afterwards, the user can change the default name of the new geodatabase. This appendix will refer to the geodatabase as "COBasins.gdb."

---

[15]New geodatabases can also be created in ArcToolbox using the "Data Management Tools" and selecting "Workspace," and then "Create File GDB."

[16]New domains can also be created in ArcToolbox in the "Data Management Tools" by choosing "Domains," and "Create Domain." Using ArcCatalog to create the domains allows the editor to view all the domains at once.

Creating Domains for the Geodatabase

Domains constrain the permissible values that can be entered into a field in an attribute table. Standardization of field values by using domains permits simpler searching, feature selection, and programming on such fields. Domains are created at the geodatabase level of organization and can be used for any attribute of any feature class in the geodatabase. Additions and edits to domains can be made only by the creator of the domain. All domains used in the COBasins.gdb geodatabase are coded-value domains (domains that allow only certain specified codes as values).

1. In ArcCatalog,[16] right click on the name of the geodatabase, select "Properties…," and select the "Domains" tab. To add a new domain, select an empty cell in the "Domain Name" column, type a name, and add a description. Two domains are required in this geodatabase: "Linesource" and "YesNo." Two other optional domains are shown in table 6 for two optional fields ("SiteType" and "GageType") in the Sites feature class.

2. In the "Domain Properties" area (middle area in the dialog), specify the field type, the domain type (coded value), the split policy, the merge policy, and the coded values and their descriptions. The split policy controls the allowable domain values for lines split during an ArcMap editing session. The merge policy controls which domain value takes precedence if features are merged.

3. Figure 3 shows the "Database Properties" window after adding the required "Linesource" and "YesNo" domains and the two optional domains, "GageType" and "SiteType." To add the coded values and descriptions, select the domain name in the uppermost section of this window. In the "Coded Values" section at the bottom of the dialog, type the coded values for each of the Domain Names as shown in table 6. In the "Description" area, type the descriptions from table 6. When viewing field entries in an attribute table, the domain "Description" is shown; when calculating new values for the domain, the contents of the "Code" are used.

Creating a Basins Feature Dataset

A feature dataset is a container for feature classes that share the same geographic extent and coordinate system. A common projection system and persistent topological relations can be set for feature classes only if they reside within a common feature dataset. New feature datasets are made using either ArcCatalog or ArcToolbox. The feature dataset will be referred to in this appendix as "BasinsFD."

Because one of the purposes of this geodatabase is to store a published numeric drainage area, the projection system should be one that is conservative with numeric drainage area

Table 6. Domains used in the basin geodatabase.

Domain name	Domain description	Field type	Domain type	Split policy	Merge policy	Domain codes and descriptions
Linesource	Source of lines used in line feature class	Text	Coded Values	Duplicate	Default Value	WBD: National Watershed Boundary Dataset; DIGITIZED: Digitized line
YesNo	Yes or no field	Text	Coded Values	Default value	Default Value	Y: Yes; N: No
GageType[1]	Type of gage	Text	Coded Values	Duplicate	Default Value	CM: Continuous-monitoring gage; CSG: Crest-stage gage or peak-flow site; SDM: Surface data measurments only; DCP: Data-collection platform at peak-flow site; NA: Not applicable
SiteType[1]	Type of site for which basin was delineated	Text	Coded Values	Duplicate	Default Value	Gage: Gage; Confluence: Confluence; WQ: Water-quality-measurement site; Gage,WQ: Both gage and water-quality-measurement site; Other: Other

[1]Optional domains used for optional fields in the Sites feature class.

values. Numeric drainage areas are calculated automatically for polygon feature classes and stored in the polygon "Shape_Area" field. To store this attribute with the highest accuracy, the feature dataset should use an equal-area projection system, such as an Albers equal-area-conic projection system. Map projections cannot portray the spherical earth on a flat map without some distortion, but equal-area projection systems preserve numeric areas of polygon features more accurately. The numeric drainage area derived from a polygon represented on a two-dimensional computer screen using an equal-area projection is proportional to the actual basin area measured on the ground,

and a numeric drainage area derived from such projections is more accurate than those stored by feature classes having other projection systems. Although numeric drainage areas derived from equal-area projections are not exactly correct, distortion is minimized between the two standard parallels used for the projection (Wiki.GIS.com, 2012). The USGS National Water Quality Assessment Program (NAWQA) Conterminous United States (CONUS) Albers projection, North American Datum of 1983, which has a central meridian of –96° longitude, can be used throughout the conterminous United States. The standard parallels used in the NAWQA CONUS Albers equal-area-conic projection system work well throughout the CONUS, and this projection makes it easy to use the basin geodatabase feature classes with the many NAWQA GIS datasets that are in this same projection system. Alternatively, an Albers equal-area-conic projection system specific to an individual state can be created by setting the central meridian to a meridian approximately in the center of the state. Doing so will make north appear vertical in ArcMap views of the state (Curtis Price, U.S. Geological Survey, written commun., 2010). If a projection system other than an equal-area-conic projection system is used, the "Calculate Area" tool in ArcToolbox and in XTools (XToolsPro, 2012) should be used periodically to recalculate numeric drainage areas. The XTools version of this tool allows one to specify a coordinate system to be used for the calculation.

Creating a new feature dataset can be done in ArcCatalog or by using tools in the ArcToolbox. The ArcCatalog method is described here:

1. Right click on the name of the geodatabase in ArcCatalog, select "New," and select "Feature Dataset." Name the feature dataset. The feature dataset is referred to in this report as "BasinsFD." Select "Next."

2. Select the projection system in this second dialog window called "New Feature Dataset" (fig. 4). In the group of projections called "USGS_Favorites," select the "Albers_CONUS_NAD83" projection.

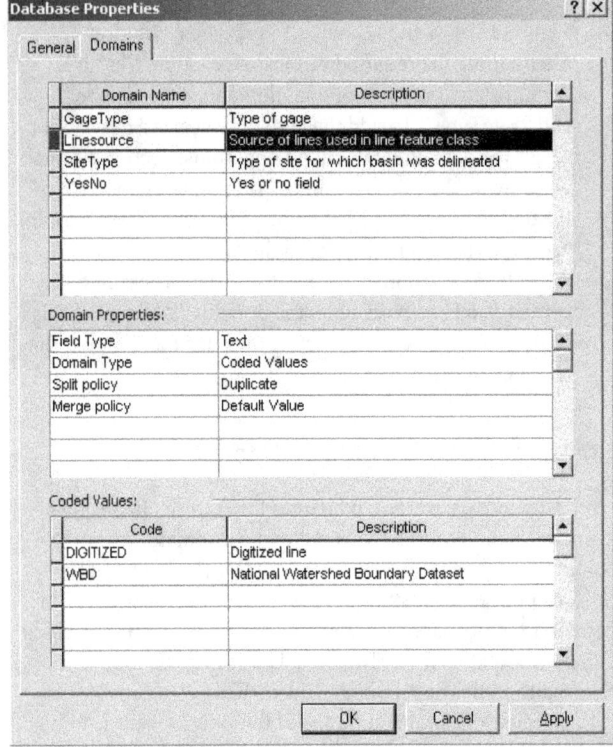

Figure 3. Domains used in the basin geodatabase.

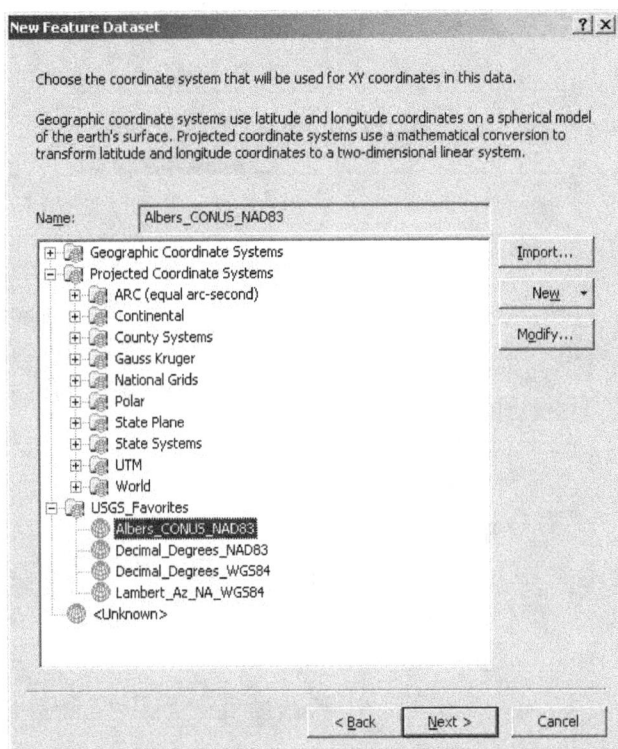

Figure 4.   Dialog in which a coordinate system is assigned during creation of a feature dataset.

3.   Alternatively, create a state-specific Albers equal-area-conic projection system using the "New" button (fig. 4), select "Projected…," name the projection, and use the listbox to pick "Albers" for the "Projection Name." First, use the "Select" button to specify a geographic coordinate system: select "North America," select the "Add" button, select "North American Datum 1983.prj," and then select "Add." After setting the coordinate system, type in the values for the new Albers equal-area-conic projection system. The "Central_Meridian" should be set to a meridian of longitude that approximately bisects the state; the "Latitude_Of_Origin" should be the latitude at the southernmost extent of the state or one just south of the state. "Standard_Parallel_1" and "Standard_Parallel_2" can be within one-half of a degree of (and internal to) the latitude parallels nearest the northernmost and southernmost state boundaries, respectively. An example of making the Colorado-specific, Albers equal-area-conic projection system used for the USGS CWSC geodatabase is shown in figure 5.

4.   Select "Next," and accept defaults for the coordinate system for "Z values" (none).

5.   Select "Next," and set the "XY tolerance" for the BasinsFD feature dataset to 0.001 meters to match the cluster tolerance used in the National Watershed Boundary Dataset (WBD). Accept the default values for the "M Tolerance" and the "Z Tolerance." Select "Finish."

Creating Feature Classes

A feature class is a table that stores information about features that share a common geometry, which may be point, line, or polygon. New, empty feature classes can be created in ArcCatalog at either the geodatabase or feature dataset level. Within the BasinsFD feature dataset are a minimum of three feature classes: a point feature class, containing site points for which polygonal basin areas are derived; a line feature class, containing all basin perimeters; and a polygon feature class, containing all polygonal basin areas. Two additional, temporary feature classes—a line and a polygon feature class—also will be created to hold features used to make the temporary, new basin feature during the delineation procedure.

Creating the Point Feature Class

Two types of site points can be stored for the basins: a type that contains the recorded location of the streamgage orifice, water-quality sampling site, confluence, or "other" site type for which the basin is delineated, or a second type that contains the true topographic outlet location for water flow exiting the basin. For the COBasins.gdb, only the first type of point feature is archived. Locations for many points of the

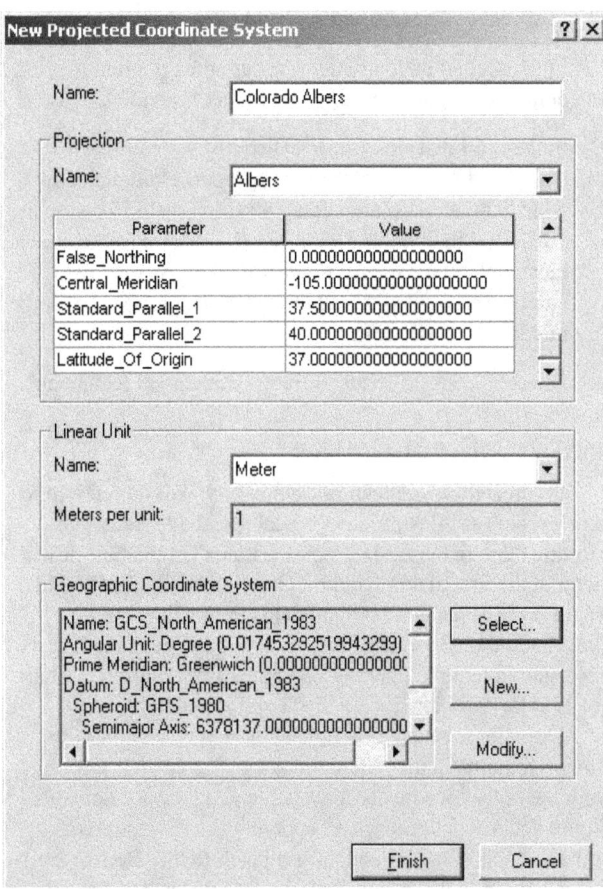

Figure 5.   Dialog in which a state-specific, Albers equal-area-conic projection system can be created.

second type, the topographic outlet location, are available as part of the high-resolution National Hydrography Dataset (NHD) as the NHDPoint feature class.

Most point locations in this geodatabase are derived from the National Water Information System (NWIS) Web site (U.S. Geological Survey, 2012a) or from a current NWIS site-file. There are instances where the site location shown on the NWISWeb location map or in the sitefile appears to be incorrect; in such instances, the site-point location is derived from the site description, from site coordinates, or from discussion with USGS or state personnel familiar with the site. Sites that are potentially mislocated are discussed with personnel in the Water Science Center Data Section to rectify the location in NWIS. Confluence and other non-NWIS sites are usually generated with reference to digital satellite imagery and/or digital topographic maps.

The following steps are used to create the point feature class that stores basin-site points:

1. In ArcCatalog, right click on the feature dataset BasinsFD and select "New" and then "Feature Class." In the first dialog window, name the point feature class, referred to in this report as "Sites." Leave the "Alias" textbox blank, and for the type of features stored in the feature class, use the pull-down menu to select "Point Features." Leave the "Geometry Properties" checkboxes in the lower window unchecked. Select "Next."

2. In the second dialog window accept the option to use a default configuration keyword. Select "Next."

3. In the third dialog window, called "New Feature Class" (fig. 6), add the field names and properties tabulated in table 7. New fields are added by typing them in below the "SHAPE" field in the "Field Name" column of the upper-most table in the dialog window (fig. 6). As new field names are added, a "Field Properties" table appears in the middle of the window.

4. Select "Finish" to create the new, empty point feature class.

Creating the Line Feature Class

To create the line feature class, which will be referred to in this report as "BasinLines," repeat the above steps, substituting "line features" for "point features" in the pull-down menu in the first dialog window. Only one additional "Field Name" will be added to default fields in this line feature class—add the "SourceOfLine" field and the field characteristics listed in table 8. There is a domain, "Linesource," which needs to be specified for this as the field is added.

Domains are added by right clicking to the right of the word "Domain" in the "Field Properties" table and selecting from the pull-down menu of available geodatabase domains. Figure 7 shows an example of applying the "Linesource" domain, created at the geodatabase level, to the "Source-OfLine" field in the BasinLines line feature class. After this domain is established for this field, only the names of GIS datasets specified in the "Linesource" domain are allowable

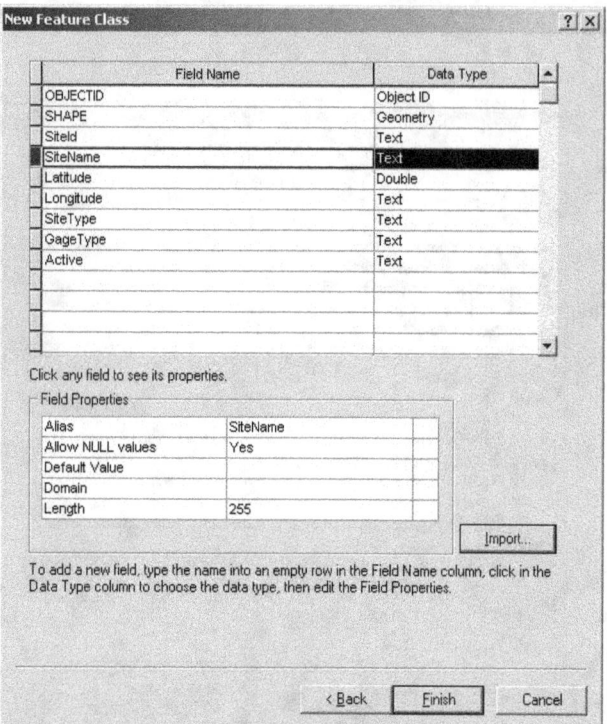

Figure 6. Dialog showing fields for the Sites feature class in the "New Feature Class" window.

Table 7. Attributes and field characteristics of added fields in the Sites feature class.

[N.A , not applicable]

Field name	Data type	Allow null values	Default value[2]	Domain	Length
SiteId	text	yes	none used	none used	15
SiteName	text	yes	none used	none used	255
Latitude	double	yes	none used	none used	N.A.
Longitude	double	yes	none used	none used	N.A.
SiteType[1]	text	yes	none used	SiteType	50
GageType[1]	text	yes	none used	GageType	5
Active[1]	text	yes	none used	YesNo	1

[1]Optional fields, shown in geodatabase.

[2]No default value needed for fields in this feature class.

entries for the "SourceOfLine" field. When editing the attribute table for this feature class, only the domain descriptions entered for the "Linesource" domain will appear in a pull-down menu of allowable values for this field.

Creating the Polygon Feature Class

Create a polygon feature class, which will be referred to in this report as "BasinPolys," by repeating the above steps, selecting "polygon features" in the pull-down menu in the first dialog window. Specify the field names and field properties as detailed in table 9. Also, apply the domain, "YesNo" to the field called "Approved" (and to the optional field called "NWISChanged") and set the default value for the domain to "N."

Table 8. Attributes and field characteristics of added fields in the BasinLines feature class.

Field name	Data type	Allow null values	Default value[1]	Domain	Length
SourceOfLine	text	yes	none used	Linesource	50

[1]No default value needed for field in this feature class.

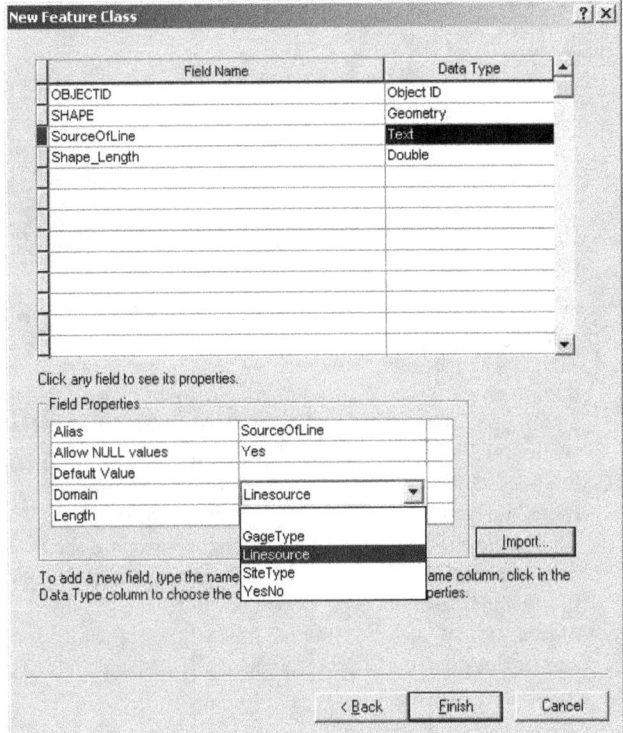

Figure 7. Dialog in which the "Linesource" domain is added for the "SourceOfLine" field.

Table 9. Attributes and field characteristics of added fields in the BasinPolys feature class.

[N.A., not applicable]

Field name	Data type	Allow null values	Default value	Domain	Length
SiteId	text	yes	none used	none used	15
SiteName	text	yes	none used	none used	255
AreaSqMi	double	yes	none used	none used	N.A.
AreaRounded	text	yes	none used	none used	6
Approved	text	yes	N	YesNo	1
NWISChanged[1]	text	yes	N	YesNo	1

[1]Optional field, shown in the geodatabase.

[17]Temporary topology, map topology, can be implemented in ArcMap editing sessions but only persists as long as the editing session. Map topology and the Map Topology tool on the Topology toolbar should not be used for feature classes if persistent topology rules created in ArcCatalog or ArcToolbox already exist for the same feature classes.

Creating Temporary Line and Polygon Feature Classes for New Basin Delineation

To store temporary new lines and/or new polygons used to create new drainage basins, two additional feature classes are needed, a line and a polygon feature class, referred to in this report as the "NewBasinLines" feature class and the "NewBasinPoly" feature class, respectively. These feature classes will have the same field names and attributes as the "BasinLines" and "BasinPolys" feature classes that were just created. Use the "Import" functionality to copy the fields and field definitions from these feature classes. Add each of these two new feature classes, and in the third dialog window that shows the field names for the new feature class, select the "Import" button to browse to and import the fields from the permanent feature classes that match geometry with the new feature class. For example, figure 8 shows the creation of the NewBasinPoly feature class. The "Import" button was used to browse to BasinPolys and to import the field definitions from BasinPolys into NewBasinPoly.

Creating Topology

Topology manages spatial relations for features within a feature class and between feature classes. For topology to be persistent in the geodatabase, it must be created in ArcCatalog or by using ArcToolbox.[17] Unlike the topology rules in ESRI coverages, geodatabase topology rules can be added or deleted. Temporary topology rules can be added occasionally

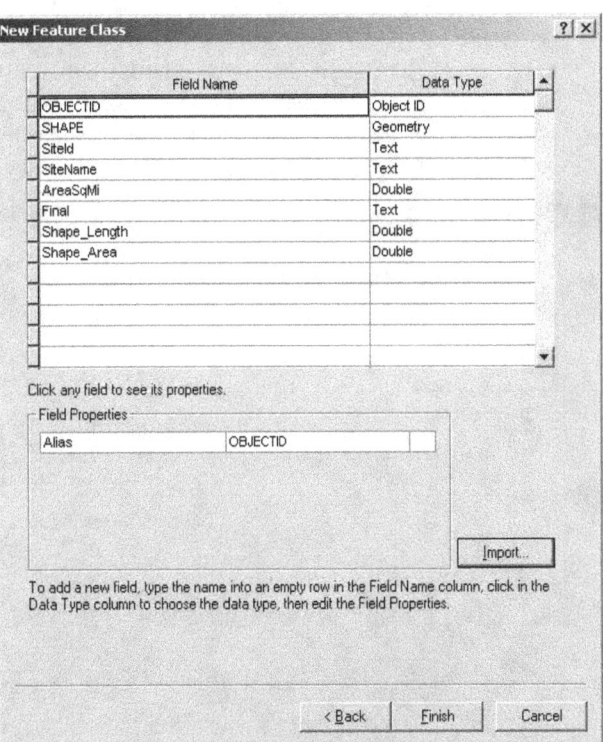

Figure 8. Dialog showing creation of the NewBasinPoly temporary feature class by importing fields from BasinPolys.

to test the spatial interrelations of the features in the geodatabase. For example, a rule can be added to ensure that site points are overlapped by basin polygon boundaries (specifically, "Sites Must Be Covered By Boundary of BasinPolys"). After the basins have been checked against this rule, the rule can be deleted. Topology rules only identify features that do not obey the topology rules and do nothing about correcting errors. Errors are corrected in an ArcMap editing session. The topological relations among the features are not stored; topology rules are applied "on-the-fly" to features in the map view as a topological selection is made.[18]

Topology for the BasinsFD feature dataset is created as follows:

1. In ArcCatalog, right click on the feature dataset; select "New," and then "Topology." Select "Next" to bypass the first dialog window.

2. In the second dialog window, type the name of the new topology, which will be referred to in this report as "BasinsFD_Topology." Accept the default cluster tolerance (which should be 0.001 meters—equal to the "XY Tolerance" set for the BasinsFD feature dataset above), and select "Next."

3. In the third dialog window, called "New Topology" (fig. 9), select those feature classes that will participate in the topology. Check BasinLines, BasinPolys, and NewBasinLines. Select "Next."

4. In the fourth dialog window, called "New Topology" (fig. 10), ranks are assigned to the feature classes in the topology. Ranks are used if features in one feature class have more reliable positions than the features in another feature class. The highest rank is 1, and feature classes of accurate data should be given a rank of 1. BasinPolys and BasinLines are intended to have a high accuracy and will be assigned a rank of 1, whereas newly added lines in the

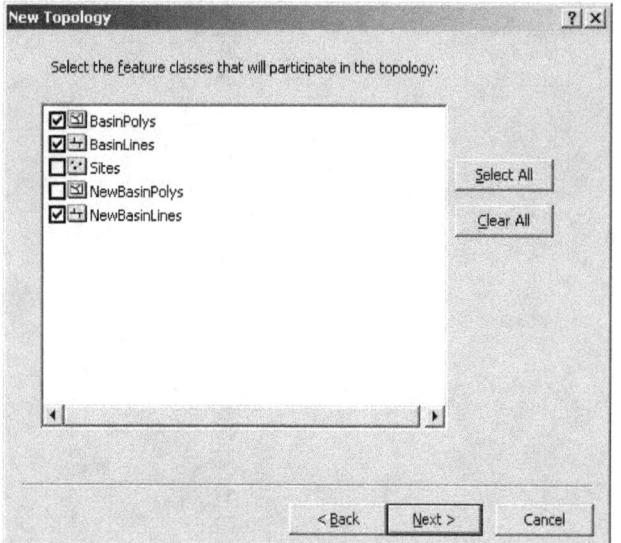

Figure 9. Dialog showing feature classes that will participate in the topology rules.

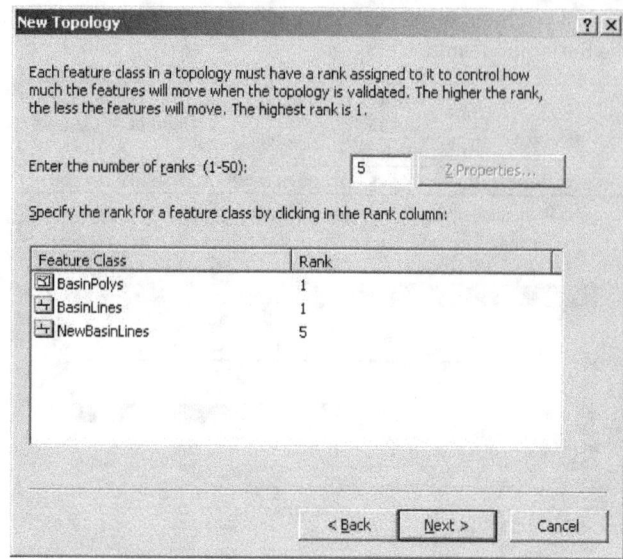

Figure 10. Dialog in which feature-class ranks are assigned during topology creation.

temporary NewBasinLines feature class will be assigned a rank of 5. When topology is validated, if feature vertices are found within the cluster tolerance, the features from the feature class with the lowest coordinate accuracy (here, the NewBasinLines with the rank of 5) will be snapped to the features with the higher coordinate accuracy (the BasinPolys and BasinLines having a rank of 1).[19] In the upper textbox, type "5" in the "Enter the number of ranks (1-50)" textbox; set the ranks of both BasinLines and BasinPolys to "1" and the rank of NewBasinLines to "5." Select "Next."

5. A fifth dialog window appears in which all topology rules will be listed. To add a new rule, select the "Add Rule" button to access a subsidiary dialog. In figure 11, the "Must Not Have Dangles" rule is being added for the BasinLines feature class.

6. Add the topology rules listed in table 10. These rules include the four basic topology rules as well as three additional temporary rules that will be used only for basin delineation. These three additional rules will control internal spatial relations for the NewBasinLines feature class and will bind spatially the features in NewBasinLines to those in BasinLines.

7. The dialog window should look like figure 12. Select "Next."

8. A sixth and final dialog window lists the topology parameters set in previous dialog windows. Select "Finish." A message may appear saying that no features were found within the extent of the topology; if so, select "OK."

---

[18]The "on-the-fly" application of topological rules is indicated by a message that appears when a feature is topologically selected, "Building topology cache on selected features."

[19]In actuality, even vertices of features ranked as 1 will shift together slightly.

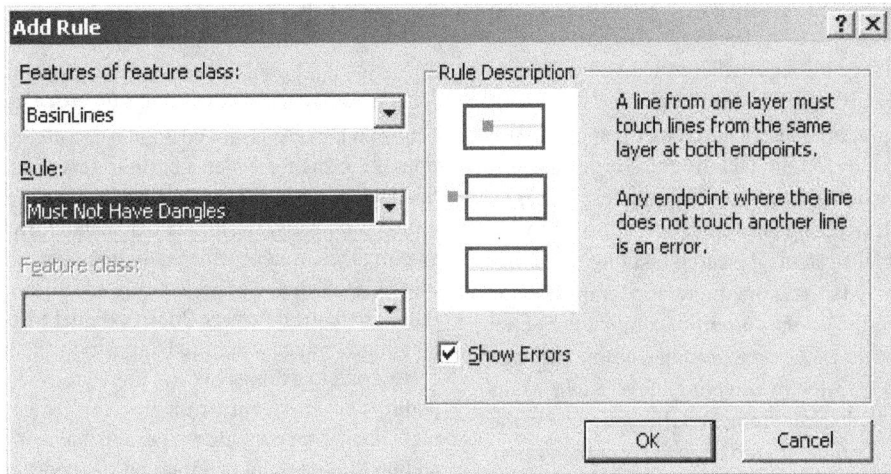

Figure 11.   Dialog in which topology rules are added.

Digital GIS Datasets Used for Basin Delineation

This second section describes several digital layers that are either essential or helpful for basin delineation. Two GIS datasets are used as a primary reference for on-screen positioning of drainage divides: the WBD, and the 1:24,000-scale USGS digital raster graphics (DRG) maps. Other useful GIS datasets include the NHD and several ESRI online map services. Sources and steps for acquiring these backdrop GIS datasets are described in this section.

National Watershed Boundary Dataset (WBD)

Prior to basin delineation, the WBD should be downloaded from the U.S. Department of Agriculture (USDA) Natural Resources Conservation Service (NRCS) Geospatial Data Gateway (*http://datagateway.nrcs.usda.gov/*). The

WBD in a Universal Transverse Mercator (UTM) projection system can be downloaded on a state-by-state basis from the GeoSpatial Data Gateway (U.S. Department of Agriculture Natural Resources Conservation Service 2012a). Alternatively, 12-digit WBD for the CONUS is available as a zipped file from a USDA NRCS ftp site (U.S. Department of Agriculture, Natural Resources Conservation Service, 2012b). This latter GIS dataset can be unzipped and clipped to a state boundary or to several selected adjacent states using ArcToolbox "Clip" tool.

In the event that changes need to be made to the WBD, the WBD steward for the state should be contacted and supplied with shapefiles containing the revised polygon and line boundaries. The names of the WBD stewards for each state are available at the USGS NHD Web site (U.S. Geological Survey, 2012c).

Table 10.   Topology rules used in the basin geodatabase.

[N.A., not applicable]

Feature class	Topology rule	Feature class
BasinLines[1]	Must Not Have Dangles[2]	N.A.
BasinLines	Must Not Intersect Or Touch Interior	N.A.
BasinLines	Must Be Covered By Boundary Of	BasinPolys
BasinPolys	Boundaries Must Be Covered By	BasinLines
NewBasinLines[3]	Must Not have Dangles	N.A.
NewBasinLines[3]	Must Not Intersect Or Touch Interior	N.A.
NewBasinLines[3]	Must Be Covered By Feature Class Of	BasinLines

[1]The topology rule is read as a sentence from left to right: "BasinLines Must Not Have Dangles."

[2]The topology rules follow the capitalization used by the Environmental Systems Research Institute (ESRI).

[3]One of three added topology rules to constrain the spatial behavior of NewBasinLines during basin delineation.

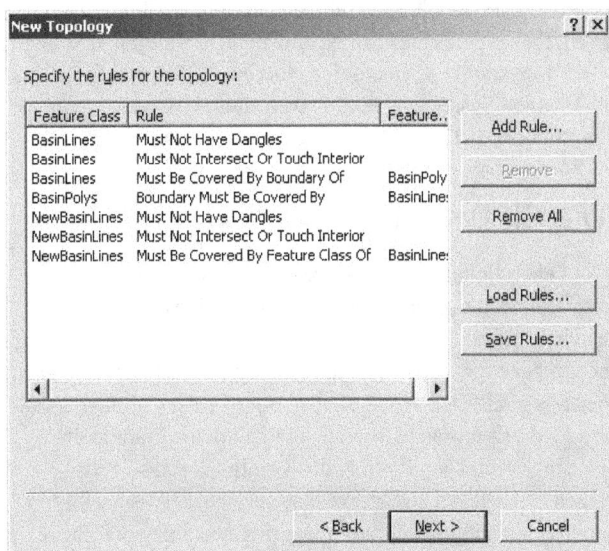

Figure 12.   Dialog showing a list of all topology rules needed for adding new basins.

USGS Digital Raster Graphics

In general, drainage basins are delineated along the topographic drainage divides using backdrop digital USGS 1:24,000-scale topographic maps. Digital topographic maps, originally published in UTM projections and North American Datum of 1927, are available from several sources, one of which is the online ESRI Map Service called "USA Topo Maps" (Environmental Systems Research Institute, 2012b). When added to an ArcMap session, digital topographic maps are projected "on-the-fly" to the current data frame projection.

The 1:24,000-scale DRG maps from the online map service can be added to ArcMap by selecting "File" in the Main menu and selecting "Add Data from ArcGIS Online." On the ESRI ArcGIS Resource Center Web page the user then can select the "USA Topo Maps" layer link and select "Open" to open the layer in ArcMap. If the map redraw rate is slow, the visibility of this layer can be turned on and off as needed in the Table of Contents window.

National Hydrography Dataset

Medium-resolution (1:100,000) and high-resolution (1:24,000) versions of USGS National Hydrography Dataset (NHD) are available for the entire conterminous United States. Medium-resolution or high-resolution NHD is used to help guide selection of WBD features for inclusion in a given basin during basin delineation. Medium- and high-resolution NHD are used during the final basin review process as the first step in checking and finalizing the polygonal basin area. In general, because the NHD is being used to pick WBD hydrologic unit codes (HUCs) that belong to the basin being delineated, the less-detailed, medium-resolution NHD is the easiest to use. The NHD is available from the USGS NHD team on a state-by-state basis. Medium-resolution and high-resolution NHD layers for several states can be downloaded from the USGS NHD Web site (U.S. Geological Survey, 2012b) and from the Horizon Systems NHDPlus Web site (Horizon Systems Corporation, 2012).

Other GIS Datasets

Other digital layers that are useful as backdrops for basin delineation include several ESRI online map services (Environmental Systems Research Institute, 2012b), including the "Imagery" map service, the "DeLorme" map service, and the "Shaded Relief" map service. Any of these GIS layers can be added to ArcMap by selecting "File" in the Main toolbar, selecting "Add Data From ArcGIS Online," and selecting "Open." The ESRI "Imagery" map service layer is useful for locating streamgages and for digitizing new linework; the ESRI "DeLorme" and ESRI "Shaded Relief" layer are both good layers to use to get an overview of the area in which a basin is to be delineated.

Creating the Basin-Site Point

This third section describes several ways to input a site-point location, which determines the delineation of the polygonal basin area. A basin-site point is the digital GIS representation of the site for which a basin is generated. Before a basin is delineated, the location of the basin-site point should be determined accurately. Basins are delineated for various site types: streamgages, water-quality sampling sites, confluences, or other USGS study- or project-related site locations. NWISWeb (U.S. Geological Survey, 2012a), digital topographic maps, and digital imagery are used to position site locations. Once the correct site location is known, there are at least three ways of adding new site-point locations to the geodatabase Sites feature class, including copying the point from another GIS dataset, adding the site point by inputting site coordinates, or adding the point interactively.

Locating the Site Point

NWIS site locations are generated from NWIS sitefile locations, NWIS sitefile coordinates, the NWISWeb location map, the site description from the Annual Water Data Report, or from discussions with USGS or state field personnel who have recently visited the site. Both confluence and NWIS site locations should be compared to the ESRI "USA Topo Maps" map service or to the ESRI "Imagery" map service (Environmental Systems Research Institute, 2012b) and to any other available information. In situations in which a stream channel has changed its location and the site point is not on a streamline shown on the historical topographic map, the site may be positioned at a global-positioning-systems- (GPS-) measured location (whether the stream channel is shown on the digital topographic map or not).

The following steps are used to ascertain the correct location of an NWIS site:

1. Access NWISWeb (*http://waterdata.usgs.gov/nwis*), (U.S. Geological Survey, 2012a) to set up a search based on site information. Enter the site number, and select the site number link to view the site-information Web page.

2. On the site-information Web page, first review the site description from the current Annual Water Data Report (if the site is active). To do this, use the pull-down list for "Available data for this site" and select the option to view "USGS Annual water-data report(s): (Offsite)." Read the site "Location" section, which is a detailed site description including measured distances from prominent local features. If the basin is being developed for an older streamgage location, refer to the USGS publications Web site (U.S. Geological Survey, 2012d) to find older Annual Water Data Reports. Search the Web site using the key words: "Water," "Data," "Report," a single year from the streamgage's period of record, and the name of the state.

3. Also, on the NWISWeb site-information Web page, in the pull-down list for "Available data for this site," there is an option to view a "Location map" (fig. 13). Selecting the

"Location map" link accesses a Google Maps view of the site area with various options for backgrounds for viewing the site location. On the "Location map" page, select either the "Satellite" view (for views of the site referenced to satellite imagery), "Hybrid" view (for views of the site referenced to satellite imagery, transportation networks, and towns), or "USGS" view (for views of the site referenced to USGS digital topographic maps), and zoom in to the site location using the plus (+) symbol.

Adding a New Site Point

New site points can be added three ways: by copying the site point from a GIS dataset made of the NWIS sitefile features, by typing in the site latitude-longitude coordinates, or interactively adding the site point by manually digitizing the site point. Use USGS DRG topographic maps as a backdrop for the ArcMap view; these can be viewed using the ESRI "USA Topo Maps" map service. Satellite imagery (available from ESRI "Imagery") is also useful for viewing the gage house and roads to the gage. After each site point is added, it should be compared to the location shown on the NWISWeb location map. The NWIS sitefile location (which should correspond to the location shown on the NWISWeb "Location Map") is used

as the primary reference for the site-point location. Site points also can be added using the site coordinates on the NWISWeb page, and this method can be a surrogate for using a GIS dataset of NWIS sitefile points. Site points can also be added manually by digitizing, a method that is slightly less exact. If the site point appears to be mislocated, discuss the site-point location with USGS personnel in the Water Science Center Data Section to determine the correct site location.

Several preliminary steps are common to all three methods:

1. Add the Sites feature class to ArcMap, and zoom to the approximate area of the streamgage or the water-quality sampling site.

2. If the Editor toolbar (fig. 14) is not displayed, add it by selecting "View" in the ArcMap Main menu, selecting "Toolbars," and checking "Editor."

3. If copying a site point from the NWIS sitefile or manually digitizing the site point, start an ArcMap editing session, select the word "Editor" on the Editor toolbar, and from the pull-down menu select "Start Editing." Use the pull-down menu for the "Edit Task" to select "Create New Feature" and use the pull-down menu for the edit "Target Layer" to select the Sites feature class as the target layer. If a message indicating that the layers to be edited are not in the map's coordinate system appears, select the "Start

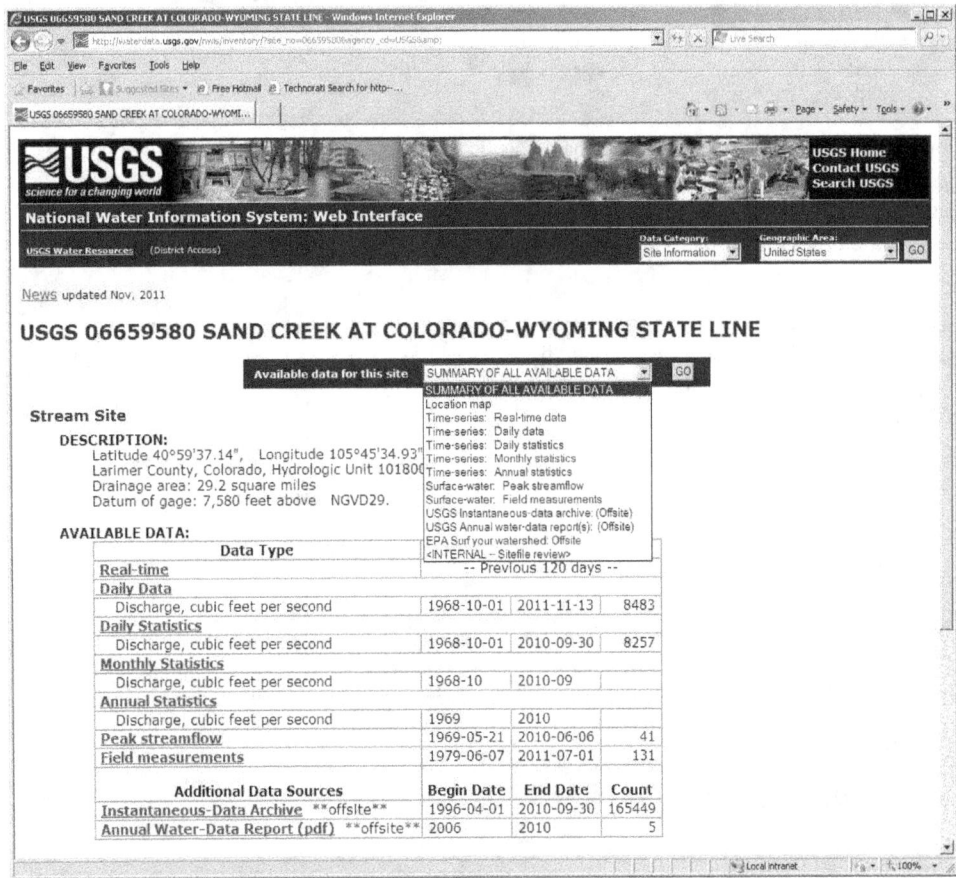

Figure 13.   National Water Information System (NWIS) Web site site-information page.

Editing" button to dismiss the message. This message can be ignored because the feature dataset in the geodatabase will force all features in it to be in a single projection system, the Albers equal-area-conic projection system. Other data layers in the ArcMap session are projected "on-the-fly" by ArcMap to a single, common projection system.

Copying the Site Point from another GIS Dataset

One of the most accurate methods of adding NWIS site points to the Sites feature class is to copy them from another GIS dataset, such as a shapefile made of the current NWIS sitefile data.

1. Download from NWIS a text file of sites (the NWIS sitefile) for a given state. The text file is added into ArcMap and will appear as a list in the ArcMap Table of Contents window when the "Source" tab is selected. Right click on the name of the text file, and select "Display XY Data…" In the "X Field" and "Y field" pull-down menus, select the longitude field and latitude field from the NWIS text file, respectively, and then select "Edit." Choose "Select," then select "Geographic Coordinate Systems," then select "North America," and select "North American Datum 1983.prj." Select "Add" to select this projection, then select "OK" and "OK." The text file is added to the Table of Contents window as an "event theme." Features in event themes are not selectable, so right click on the event theme, and select "Data," and then "Export Data…," to export the event theme as a shapefile.

2. Add this shapefile of NWIS sites to ArcMap, and make this source GIS dataset the only selectable layer by selecting the "Selection" tab at the bottom of the Table of Contents window and checking only the checkbox for the source GIS dataset.

3. In an ArcMap editing session, make the edit "Target Layer" is the Sites feature class, and select the new site point from the shapefile by drawing a small selection box around it using the "Edit Tool" (fig. 14) on the Editor toolbar.

4. From the main menu select "Edit," click "Copy," and then select "Edit," and click "Paste."

5. Review the site-point location using the guidelines in the section on "Locating the Site Point," above; compare the location both to the site description in the Annual Water Data Report and to the map view of the site on NWISWeb (U.S. Geological Survey, 2012a). If the site appears to be mislocated, its location should be confirmed by other USGS personnel in the Water Science Center Data Section.

6. With the new site point selected, update the attributes in the Sites attribute table by selecting the "Attributes" button on the Editor toolbar (fig. 14) to open the attributes window (fig. 15). To avoid typing errors, it is best, if possible, to copy and paste attributes from the source shapefile of site locations. Populate the "SiteId," "SiteName," "Latitude," and "Longitude" fields, and, if present, the optional "SiteType," "GageType," and "Active" fields.

7. Save edits and continue on to the section "Delineating Basins."

Adding the Site Point Using Site Latitude-Longitude Coordinates

A second method of adding a new site-point location to the Sites feature class is to input site coordinates in decimal degrees, which may be derived from the coordinates displayed on the NWISWeb site-information Web page (U.S. Geological Survey, 2012a). Site coordinates are available on the site-information Web page in two places, both accessible from the pull-down menu for "Available data for this site" on the site page. Site coordinates in degrees, minutes, and seconds are shown on the "SUMMARY OF ALL AVAILABLE DATA" (default site-information Web page); site coordinates in decimal degrees are available on the "<INTERNAL – Sitefile review>" page, which is only accessible to USGS employees. Using these NWIS site coordinates may serve as an alternative to deriving a sitefile from NWIS or as a check to the location shown on the NWISWeb location map.

1. Input coordinate values should be in decimal degrees – if using the coordinates given in degrees, minutes, and seconds displayed on the "SUMMARY OF ALL AVAILABLE DATA" NWISWeb site page, first convert the coordinates to decimal degrees.

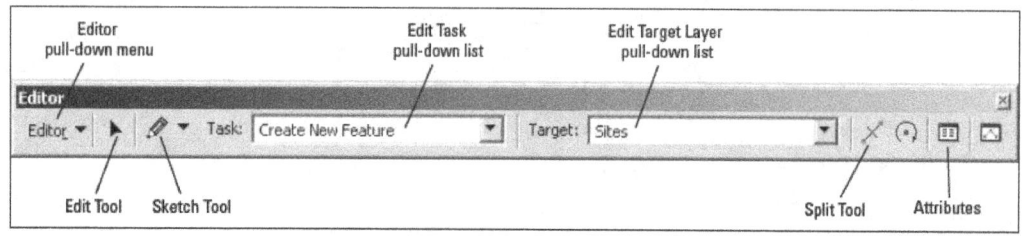

Figure 14.   The ArcMap Editor toolbar.

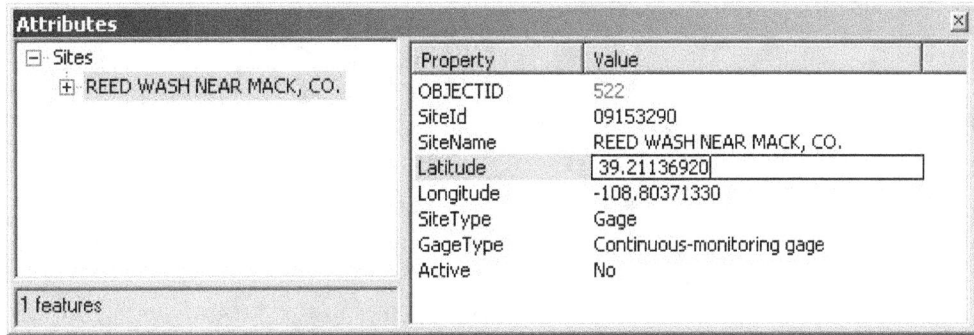

Figure 15. Attributes window accessed by selecting the "Attributes" button on the Editor toolbar.

2. Before starting an ArcMap editing session, change the coordinate system of the data frame to be a geographic coordinate system. Right click on the map view area of the ArcMap window; select "Data Frame Properties…," select the "Coordinate System" tab, select "Predefined," select "Geographic Coordinate Systems," pick "North America," select "North American Datum 1983," and click "OK." If a message appears that the coordinate system is different from one or more of the data sources in the map, click "Yes" to dismiss the warning message. This message can be ignored because the feature dataset in the geodatabase will force all features in it to be in a single projection. Any other data layers in the ArcMap session are projected "on-the-fly" by ArcMap to a single, common coordinate system.

3. Right click again in the map view, select "Data Frame Properties…," select the "General" tab, and change the "Units" for the "Display" to "Decimal Degrees" from the pull-down menu of options. If a message appears that the coordinate system is different from one or more of the data sources in the map, click "Yes" to dismiss the warning message.

4. Start an ArcMap editing session to edit the geodatabase by selecting the word "Editor" in the Editor toolbar (fig. 14, and top of fig. 16), and in the pull-down menu select "Start Editing." Use the pull-down menu for the "Edit Task" to select "Create New Feature" and use the pull-down menu for the edit "Target Layer" to select the Sites feature class.

5. Zoom near to where the site point will be added, using the digital USGS DRG maps as a guide.

6. To enter coordinate values, select the pencil-like "Sketch Tool" (fig. 14) on the Editor toolbar.

7. Right click anywhere in the map view, select "Absolute X, Y," and enter the coordinates in decimal degrees for the site-point location (X is longitude and Y is latitude; recall that longitudes in the Western Hemisphere are negative

numbers). Precision of the entered coordinates should be at least five decimal places to record positions accurately to the second (fig. 16). Click the "Enter" key on the keyboard to dismiss the dialog

8. A new site point should appear in the Sites feature class. The point will be selected and shown in the default cyan selection color. If necessary, zoom to the site by right clicking "Selection" and "Zoom to Selected Features."

9. At this stage, the site-point location shown in the ArcMap view should be compared to that on the NWISWeb location map and the site description. If there are discrepancies, the site location should be resolved using the guidelines specified in the "Locating the Site Point" section.

10. With the new site point selected, update the Sites attribute table by selecting the "Attributes" button on the Editor toolbar (fig. 14).

11. Manually type the "SiteId" and "SiteName" attributes in the attributes window (fig. 15). Also, populate the "Latitude" and "Longitude" and populate the optional "SiteType," "GageType," and "Active" text fields if they were added to the Sites feature class.

12. Save edits and exit the editing session by selecting "Editor" in the Editor toolbar (fig. 14) and selecting "Stop Editing" from the pull-down menu. Click "Yes" if prompted to save your edits.

13. Return the data frame to the Albers equal-area-conic projection system by right clicking on the data frame (map view) window, select "Properties…," click "Coordinate System," click "Import," browse to the geodatabase feature dataset BasinsFD, select the feature dataset name, click "Add," and click "OK." If warned that the coordinate system is different from layers in the map, dismiss the warning message by selecting "Yes."

14. Continue to the next section on "Delineating Basins."

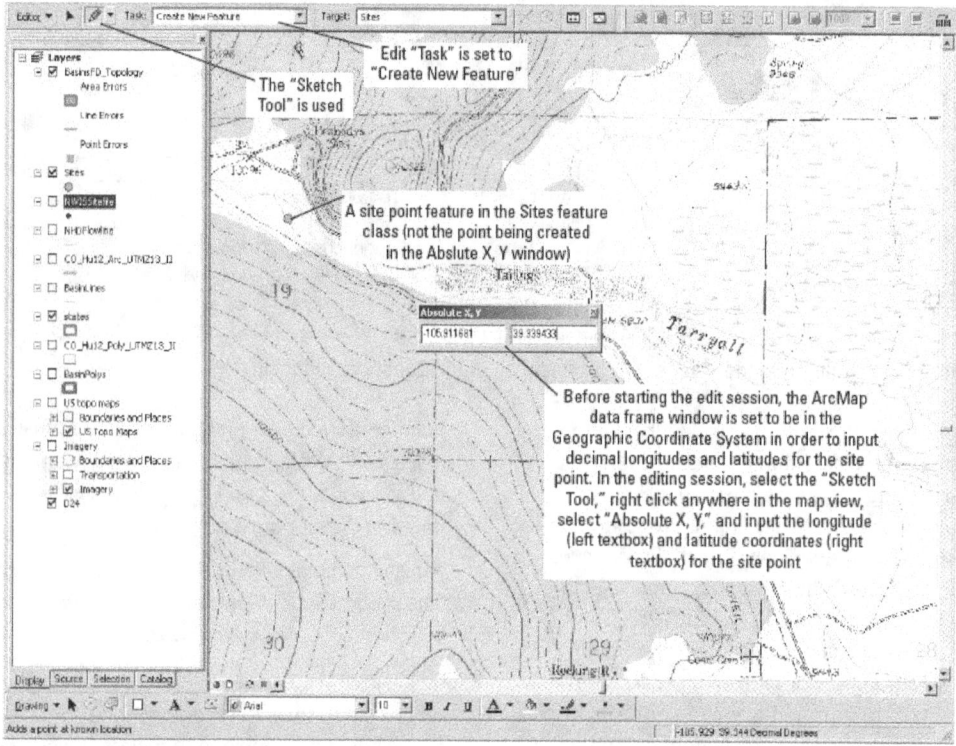

Base from U.S. Geological Survey digital data, 2012, 1:24,000
Geographic Coordinate Sytem

Figure 16.   ArcMap view of adding a new site point by manually typing in coordinates in the Absolute X,Y window.

Interactively Adding the Site Point

The third way to add a new site point to the Sites feature class is to digitize it manually. This is generally done by comparing the view in the ArcMap session to the view of the site location on the NWISWeb site location map (U.S. Geological Survey, 2012a) and to the site description from the USGS Annual Water Data Report. If the site appears to be located incorrectly in NWIS, this is a way to add the site using the site description, digital topographic maps, and satellite imagery as references.

1. On the NWISWeb site-information Web page, use the location map and the site description (if the site is active) as guidance for locating the site point. If the site is inactive, compare the site location on the NWISWeb site location map to the written description in an Annual Water Data Report for one of the years in the site period-of-record (U.S. Geological Survey, 2012d) given on the "SUMMARY OF ALL AVAILABLE DATA" site-information Web page.

2. In an ArcMap editing session, ensure that the ESRI "Imagery" and the ESRI "USA Topo Maps" map services have been added and are visible. Zoom in to the site location in both the ArcMap data frame and on the NWIS "Location map" page to compare the location.

3. Click on the "Sketch Tool" (fig. 14). Ensure that the edit "Target Layer" is the Sites feature class. Using the NWISWeb "Location map" and the backdrop DRG maps as a guide, click in the map frame to digitize the new site point. If the location on the NWISWeb "Location map" seems to be incorrect, use the site description in an Annual Water Data Report to digitize a temporary location for the site.

4. Review the site-point location using the steps in the "Locate the Site" section. If the site appears to be incorrectly located and a temporary point was digitized, its location should be confirmed with the USGS personnel in the Water Science Center Data Section familiar with the site before delineating a basin.

5. After manually adding the site point, update the "SiteId," "SiteName," "Latitude," and "Longitude" attributes in the Site attribute table by selecting the "Attributes" button on the Editor toolbar (fig. 14) to open the attributes window (fig. 15). If the optional "SiteType," "GageType," and "Active" fields were added, also populate those fields.

6. Save edits and continue to the "Delineating Basins" section.

Delineating Basins

This fourth section presents information for actual drainage-basin delineation. Before embarking on delineating a basin it is helpful to get a general picture of where the basin is in geographic space, especially with reference to digital stream networks. Two methods for delineating basins are presented here: a method that uses line features to develop a basin perimeter and then construct a polygon from that perimeter, and a second method that aggregates polygons into one feature and cuts that feature into the desired polygonal basin area. Guidelines for basin delineation and for basin finalization are also presented in this section.

Determining the Location of the Drainage Basin

It is important to try to form a mental image of the entire drainage basin before digitally capturing the features that form the basin extent. There are several ways to gain an overview of the general location of the drainage basin, including using medium-resolution or high-resolution NHD, using a query on WBD polygon attributes, or using digital elevation models (DEM) or StreamStats to generate a guide basin.

Medium-resolution NHD is one of the most useful GIS datasets for guiding the selection of features from the WBD. Before a basin is delineated, the pattern of the stream network in this GIS dataset should be visually traced upstream from the site point to determine which lines or polygons from the WBD enclose the network of stream lines draining to the site.

There are several selection methods that also can help when selecting WBD basin polygons by basin attributes. For example, the "Select By Attributes" window in ArcMap can be used to find "HUC_10_Name" field entries in the WBD that pertain to the streamgage location. For example, a Structured Query Language (SQL) statement "where clause" such as "[HUC_10_Name] LIKE 'Ohio Cr*'" used in the Colorado WBD would find all 10-digit HUC basins for Ohio Creek in Colorado. The SQL statement can also be structured to select all WBD polygons in a particular 8- or 10-digit HUC. The NHD stream networks can be useful in discerning which WBD features are missing from such selections.

If StreamStats is available for the state (U.S. Geological Survey, 2012e), one can use StreamStats to delineate a guide basin from underlying DEM data on the StreamStats Web site and import the basin into ArcMap. After this basin is made on the StreamStats Web site, click on the "Selection" menu and select "Select By Location," and select from the WBD polygons all those having centroids within the StreamStats basin.

Adding a New Basin

After the site point has been added and verified, there are a variety of approaches to delineate a basin. Two methods will be described: a basin-perimeter delineation method in which a temporary linear perimeter is made and the perimeter is used to construct a basin polygon; and a cut-polygon delineation method, in which all WBD polygons that include the basin are combined into a temporary single polygon and then cut using a single line at the downstream end of the basin. After the new features are added to the geodatabase archive using either of these methods, the topology is checked for errors, and corrections are made if there are any errors. Later, the basin is formally reviewed and finalized by comparing its polygonal basin area to the NHD stream network and by comparing its numeric drainage area to the published NWIS numeric drainage area (U.S. Geological Survey, 2012a) for the site and for upstream and downstream sites.

Before discussing the two delineation methods, initial editing steps, general digitizing steps, general topological editing guidelines, and post-delineation steps common to both methods will be reviewed.

Initial Editing Steps

Several initial steps are common to both the basin-perimeter delineation method and to the cut-polygon delineation method:

1. Before starting an editing session, check the topology to be sure it has been validated with no errors. Generally this will have been done at the end of any previous editing session. To validate the topology, open ArcCatalog; browse to the geodatabase and to the topology. Click on BasinsFD_Topology, and select the "Preview" tab in the ArcCatalog view window. If there are diagonal lines in the preview window for the topology, the topology is termed "dirty" and needs to be validated. To validate the topology, right click on the topology BasinsFD_Topology under the BasinsFD feature dataset, and select "Validate." Afterwards, right click again on the topology, select "Properties…", select the "Errors" tab, and select "Generate Summary." There should be no errors in any error category before additions are made to the geodatabase. If there are topological errors, they should be corrected before any additions are made to the geodatabase.

2. Before creating any new features, make a copy of the geodatabase. To do this, right click on the geodatabase name in ArcCatalog, select "Copy" and then click "Paste." Rename the original with a date suffix. Use the copy of the geodatabase for the editing session.

3. Open ArcMap and add the BasinsFD_Topology. When prompted by a message asking if participating feature classes should also be added, select "Yes." The BasinLines, BasinPolys, and NewBasinLines feature classes all will be added to the ArcMap session. Add also the Sites feature class if it is not in the ArcMap view. Also add the WBD HUC12 lines, the NHD medium-resolution flowlines, and the ESRI "USA Topo Maps" map service. For instructions regarding adding these layers, refer to the "Digital GIS Datasets Used for Basin Delineation" section.

4. In the ArcMap window, ensure that the Editor (fig. 14), Standard (fig. 17), Tools (fig. 18), and Topology toolbars (fig. 19) are visible: click on "View," select "Toolbars," and check these four toolbars in the toolbar list.

Figure 17.   The ArcMap Standard toolbar.

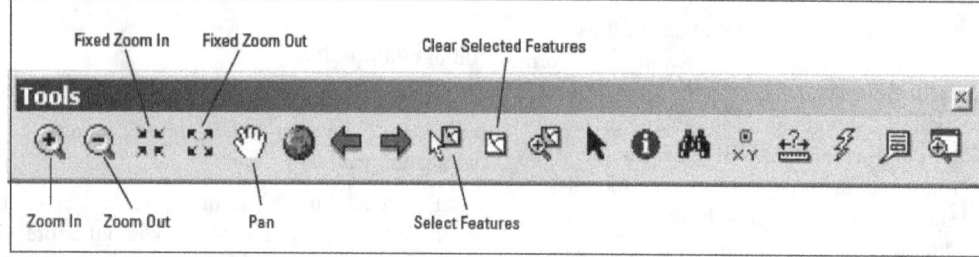

Figure 18.   The ArcMap Tools toolbar.

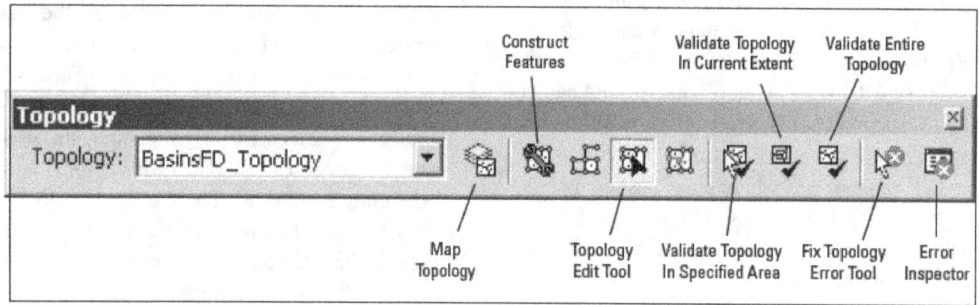

Figure 19.   The ArcMap Topology toolbar.

General Digitizing Steps

Both methods will involve digitizing two lines from the site point to the nearest feature already in the BasinLines feature class or to a feature in the WBD. Each line digitized from the site point should be the topographic drainage divide where water will flow one way on one side of the digitized line and the other direction on the other side of the line. Digitizing of the topographic drainage divide always should start at the basin-site point and work up the contours (in other words, move from lower elevations to higher). Lines for the drainage divide should be digitized to be perpendicular (as much as possible) to the topographic contour lines. Other digitizing guidelines, especially for topographic interpretation around anthropogenic and non-anthropogenic features, are given in Appendix 2.

1. In ArcMap, before starting an editing session, make the Sites feature class visible in the Table of Contents window. Also ensure that the data frame projection system is the Albers equal-area-conic projection system of the BasinsFD so that the map units are meters. To set the data frame projection, right click on the data frame (map view) window, select "Properties…", click "Coordinate System," select "Import," browse to the geodatabase feature dataset, click on the feature dataset name, click "Add," and click "OK."

2. Initiate an ArcMap editing session by selecting the word "Editor" in the Editor toolbar (fig. 14), and in the pull-down menu select "Start Editing." Select the name of the geodatabase to edit and click "OK." If a message appears indicating that the layers to be edited are not in the map's coordinate system, click the "Start Editing" button to ignore the message. This message can be ignored because the feature dataset in the geodatabase will force all features in it to be in a single projection. All data layers in the ArcMap session are projected "on-the-fly" by ArcMap to a single, common projection. The 'forced' projection system engendered by the feature dataset also ensures that all polygon areas will

be recorded in the polygon attribute table as accurately as possible (in an Albers equal-area-conic-projection-system of measurement), in spite of multiple projection systems of the different data layers being used.

3. Set the "Edit Task" in the Editor toolbar to "Create New Feature," and make the NewBasinLines feature class the edit "Target Layer" (fig. 14). In the Table of Contents window, check the 1:24,000-scale DRG topographic map GIS dataset to make it visible. Make both the BasinLines feature class and the WBD lines visible.

4. Begin digitizing a new line in the NewBasinLines feature class by snapping the first digitized vertex to the basin-site point. To make the first digitized vertex snap to this point, select "Editor" on the Editor toolbar (fig. 14); click "Snapping…" from the pull-down menu options to open the "Snapping Environment" window; and, in the top "Layer" window, check the checkbox for the "Vertex" of the Sites layer. Also set a snapping tolerance (equivalent to a snapping radius) by selecting "Editor" on the Editor toolbar and selecting "Options…" from the pull-down menu to open the "Editing Options" window. Set the "Snapping tolerance" in this window to be about 25 map units (which are meters). Click the "Sketch Tool" on the Editor toolbar (fig. 14), and then click on the basin-site point (the first vertex should snap to the site point if the "Snapping tolerance" has been set as described above. Click to add successive vertices to digitize a line from the site point along the topographic drainage divide.

5. Digitize new lines as accurately as possible using 1:24,000-scale DRG topographic maps as a backdrop. Optimally, digitizing of the topographic drainage divide should occur at map scales ranging from 1:6,000 to 1:10,000. However, periodically zooming out to at least 1:24,000 is helpful in interpreting topographic contours and in visualizing the drainage divide.

6. An example of a digitized line is shown in figure 20; each vertex created with a click of the mouse is symbolized (by the default, ArcMap symbolization settings) as a green square, and the last-digitized vertex is depicted as a red square. As the digitized line approaches the nearest feature in BasinLines, WBD, or (depending on the delineation method used) the polygon boundary in NewBasinPoly, check the checkbox for the "Vertex" of that feature class in the "Snapping Environment" window and try to snap the last digitized vertex to an existing vertex in that feature class. If no existing vertex is appropriate for the topographic divide, check the checkbox for snapping to the "Edge" to snap to a line (called an "edge") instead of to a vertex.[20] Click to digitize the last point, right click, and select "Finish Sketch" to stop digitizing.

[20]The snapping environment and other snapping options can be changed as the line is being digitized. After the snapping environment or options have been changed, click again on the "Sketch Tool" to continue digitizing.

7. If mistakes are made during digitizing, "control-z" is a keyboard shortcut to delete the last vertex in the line being digitized. Right clicking on the map view during digitizing and selecting "Delete Sketch" will delete the entire digitized line. If the line just finished needs to be deleted and re-digitized, click the "Undo" button (fig. 17) on the Standard toolbar, which moves one editing step at a time backwards to the last saved edits.

8. The ArcMap Tools toolbar (fig. 18) zoom and pan buttons also can be used during digitizing. If one of these buttons is used, simply select the "Sketch Tool" (fig. 14) again to resume digitizing. Several keyboard shortcuts also are useful. Pressing the "c" key while digitizing allows you to pan; pressing "z" will let you zoom in; and pressing "x" lets you zoom out.

9. If lines in the NewBasinLines feature class need to be corrected after they have been digitized, use either the "Modify Feature" or the "Reshape Feature" "Edit Task." The "Modify Feature" "Edit Task" is used to adjust individual vertices; the "Reshape Feature" "Edit Task" is used to modify sections of a line. To edit using the "Modify Feature" task, make only the NewBasinLines feature class selectable (use the Selection tab in the Table of Contents window and check the checkbox by this feature class) and select the line using the "Edit Tool" (fig. 14). Once a line is selected and the "Modify Feature" "Edit Task" is used, the line vertices will appear as a string of green square symbols. Using the "Edit Tool," move the cursor to hover over an individual vertex until the cursor changes to a diamond shape, and then click and drag the vertex to a new location. If the cursor is not the diamond-shaped symbol, the entire line will shift, and the "Undo" button must be used to step back one editing step. To use the "Reshape Feature" "Edit Task," make only the NewBasinLines feature class selectable, and select the line that needs to be corrected using the "Edit Tool." Click on the "Sketch Tool" (fig. 14), and click to add a vertex on one side of the line at the start of the length to be corrected. Add the second vertex by crossing the old line and continue to digitize new vertices for the new line location. When finished, cross the old line a second time before adding the final vertex. After adding the final vertex, right click and select "Finish Sketch." The line will be redrawn between the two locations where the reshaped line crossed the old line.

10. Alternatively, lines can be completely deleted by selecting them and using the "Delete" button (fig. 17) on the Standard toolbar.

11. In areas of subtle relief, it can be helpful to digitize a basic boundary line at a scale of 1:24,000 (or smaller scale) and finalize it locally afterwards using the "Reshape Feature" "Edit Task" to refine the line location.

12. Features can be selected using either the "Edit Tool" on the Editor toolbar (fig. 14) or the "Select Features" tool on the Tools toolbar (fig. 18). In either case, when selecting

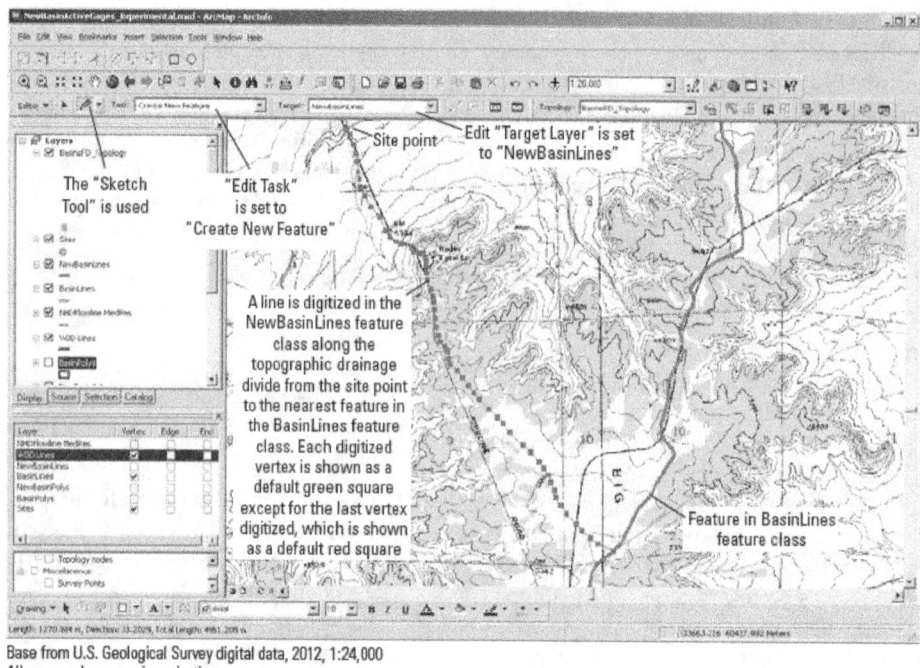

Base from U.S. Geological Survey digital data, 2012, 1:24,000
Albers equal-area conic projection
Standard parallels 37° 30' N and 40° 30' N
Central meridian 105° 00' W

Figure 20.   ArcMap view showing a newly digitized line in the NewBasinLines feature class.

point or line features for modification, it is better to draw a small selection box around the feature than to select by clicking on the feature. Clicking on the feature can occasionally displace the feature, which may not be apparent until a number of errors appear during the topology-checking stage of the workflow. Polygon features can be selected by drawing a box on any part of the feature, as long as it is checked in the Selection tab of the Table of Contents as selectable.

General Topological Editing Guidelines

There are several general guidelines for having a problem-free editing session. The Topology toolbar (fig. 19) contains the tools used for validating features and examining them for topological errors.

1. Generally, use the "Validate Entire Topology" button on the Topology toolbar (fig. 19) to validate topology. Using the other topology validation buttons on the Topology toolbar[21] to only partially validate topology without viewing and validating the entire extent of the topologically "dirty area" (the edited area where topological relations are affected) potentially allows undetected topological errors to be saved, that, if serious, can be difficult to repair once they are saved. By design and for performance reasons, topology validation has a maximum validation limit of 50,000 errors, and if the dirty area is large and contains many nested basins, validation of the entire topology occasionally will take a long time, or rarely it will not validate. If validating the entire

topology takes longer than 10 or 15 minutes, either wait or opt to cancel the validation of the entire topology, and alternatively validate portions of the topologically dirty area until it is entirely eliminated. To view the dirty area in ArcMap, right click on the BasinFD_Topology, click "Properties…", click on the "Symbology" tab, and check the box by "Dirty Areas," and the dirty area should appear as a blue hachured area in the map view. Zoom out to view the full dirty area, and use the "Validate Topology in Specified Area" button on the Topology toolbar (fig. 19) to draw a box using the mouse cursor around sections of the dirty area, and validate sections of the dirty area until no more topologically dirty areas remain. After validating each section, check for errors in the "Error Inspector" window, which is discussed in the next step. The appropriate correction for each error depends on the delineation method, and such corrections are described in the two basin-delineation-method sections below.

2. Validating only finds errors, and it does not make errors go away. Check for errors after validating by using the "Error Inspector" button (fig. 19). Make the "Error Inspector" window large enough to view the "Rule Type," "Class1," "Class2," and "Shape" fields. The "Error Inspector" window has a tendency to dock (to become locked in the toolbar area above the map view or beside the Table of Contents view), thereby decreasing the size of the map-view pane. To prevent it from docking (or to

---

[21]The "Validate Topology in Specified Area" and the "Validate Topology in Current Extent" buttons.

move it if it has docked), use the control key and click on the "Error Inspector" window to drag the window to a better location to simultaneously view both the list of errors and the map. If possible, place the "Error Inspector" window on a second monitor. When opening the "Error Inspector" window, always uncheck the "Visible Extent only" checkbox so all errors are visible, and click the "Search Now" button to view all errors.

3. Topology errors are read as a sentence from fields in the "Error Inspector" window, beginning with the "Class1" field, then the "Rule Type," any entry in the "Class2," and then the "Shape" field. For example, a "'NewBasinLines (Class1 field) Must Not Have Dangles (Rule Type field) Point (Shape field)" is the way the error should be understood.

4. Errors are displayed as a default pinkish color in the map view. When selected, errors are displayed as a default black color. Errors can be selected and viewed in two ways. Selecting an error listed in the "Error Inspector" window will highlight it as a black-colored selection in the map view. Alternatively, selecting the "Fix Topology Error Tool" (fig. 19) and selecting an error in the map will select the corresponding error in the "Error Inspector" window. The appropriate fixes for errors encountered depend on the delineation method and are discussed in the sections describing each delineation method.

5. Validate frequently, particularly after adding each new basin. This allows early detection of errors while it is still possible to use the "Undo" button to back out of mistakes. Generally, editing will be problem free. Numerous topology errors of different types generally mean that a line or polygon has been inadvertently offset. If this is the case, use the "Undo" button (fig. 17) to step back through the edits until you can validate without seeing errors in the "Error Inspector" window. If the cursor is allowed to hover over the "Undo" button, a "tip" bubble that indicates the previous step will appear. A tip bubble reading "Undo Move" means that either a vertex or node was moved (a legitimate edit) or a whole line was accidentally moved. Undo the "Undo Move" edits until there are no errors in the Error Inspector dialog window.

6. Generally, unless the potential errors are easy to repair, do not save until the topology can be validated without errors. If necessary, end the edit session, go back to an earlier version of the geodatabase, and redelineate the basin.

Post-Delineation Steps

After the basin is delineated, attribute fields should be updated, including the "SiteId," "SiteName," and numeric drainage-area attributes in the BasinPolys feature class, and the "SourceOfLine" field in the BasinLines feature class. If the "SiteId" in BasinPolys is selected, and the records are sorted in an ascending way, the new basin polygon with blank attribute values will be at the top of the attribute table.

The "AreaSqMi" field is updated using the numeric drainage area in the "Shape_Area" field (in units of square meters) by dividing the "Shape_Area" field entries by a conversion factor of $2,589,988$ m^2/mi^2. The "AreaRounded" field is updated by rounding the entry in "AreaSqMi" to NWIS specifications. The USGS uses "banker's rounding," in which a number ending in "5" is rounded to the nearest even number. If the numeric drainage area in the "AreaSqMi" field is less than 10 mi^2, the area is rounded to two decimal places. If the numeric drainage area is greater than or equal to 10 mi^2 and less than 1,000 mi^2, the area is rounded to three significant figures; if the basin numeric drainage area in the "AreaSqMi" field is 1,000 mi^2 or greater, the area is rounded to the nearest mile (Novak, 1985). For BasinPolys, the "Approved" field and the optional "NWISChanged" field should be set to "No." In the BasinLines feature class, sort ascending on the "SourceOfLine" field to find new, blank entries, and edit these to indicate if the line was derived from the WBD lines or from digitizing.

Basin-Perimeter Delineation Method

Using the following approach, a closed basin perimeter is created in the NewBasinLines feature class and the perimeter is used to construct a new polygon feature in the BasinPolys feature class (fig. 21). The perimeter is made by first digitizing two new lines in NewBasinLines from the basin-site point to the nearest line features in BasinLines or in the WBD, copying any lines already in the BasinLines feature class that are part of the new basin perimeter, copying any WBD lines that are also part of the basin perimeter, and trimming the copied lines until all lines form a single, closed loop. A "NewBasinLines Must Not Have Dangles Point" topology rule then is used to test the loop in NewBasinLines to make sure it has no missing segments. Once a closed perimeter loop is created in NewBasinLines, it is used to construct a polygon feature in BasinPolys, and the "NewBasinLines Must Be Covered By Feature Class Of BasinLines Polyline" topology rule is used to find any new lines in the NewBasinLines perimeter loop that need to be added to BasinLines.

1. Make at least the following data layers visible using checkboxes in the Table of Contents window: the WBD lines, BasinLines, NewBasinLines, and the DRG topographic maps.

2. Using 1:24,000-scale DRG maps as a backdrop, the first step is to digitize two lines from the basin-site point. Two lines are digitized on either side of the site point to the nearest features either in BasinLines or in the WBD, and both layers should be made visible during the digitizing process. If the basin is small, however, this digitizing step may entail digitizing the entire basin perimeter. The guidelines and digitizing steps in the "Initial Editing Steps" and in the "General Digitizing Steps" sections should be consulted at this stage. Both BasinLines and the WBD lines should be kept visible during digitizing because the digitized line may need to be terminated at a feature in one of these GIS datasets.

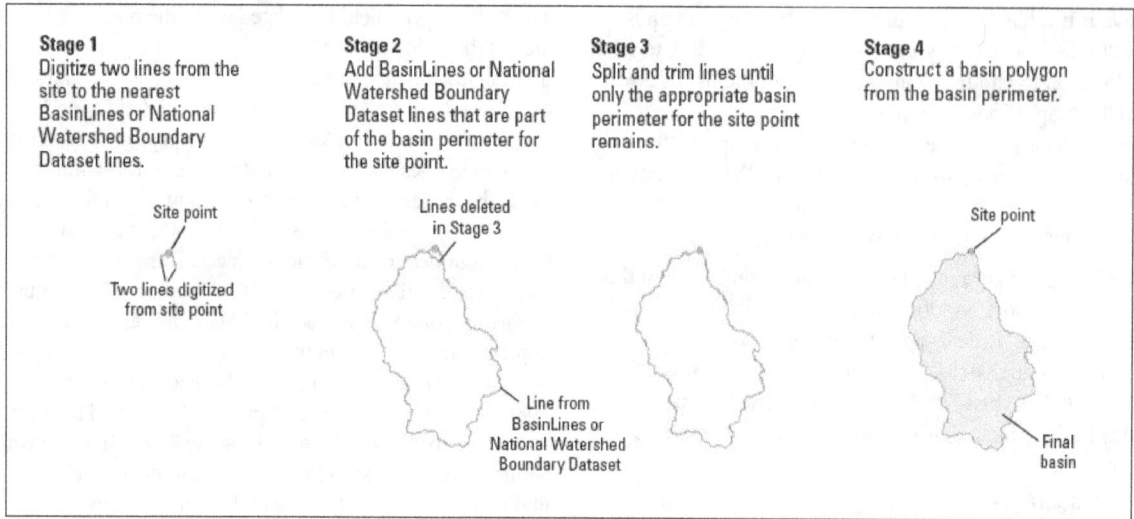

Stage 1
Digitize two lines from the site to the nearest BasinLines or National Watershed Boundary Dataset lines.

Site point

Two lines digitized from site point

Stage 2
Add BasinLines or National Watershed Boundary Dataset lines that are part of the basin perimeter for the site point.

Lines deleted in Stage 3

Line from BasinLines or National Watershed Boundary Dataset

Stage 3
Split and trim lines until only the appropriate basin perimeter for the site point remains.

Stage 4
Construct a basin polygon from the basin perimeter.

Site point

Final basin

Figure 21.   Schematic of the four stages of basin delineation using the basin-perimeter delineation method.

3. After the two lines have been digitized, the remaining lines that form the basin perimeter are added to New-BasinLines. Both BasinLines and the WBD lines should be visible. Make the medium-resolution NHD visible as an aid in selecting the lines that form the basin perimeter. Turn off the visibility of the DRG maps layer, and zoom out to view the full extent of the NHD stream pattern that drains to the basin-site point.

4. If there are any lines in BasinLines that are part of the new basin perimeter, select the "Selection" tab below the ArcMap Table of Contents window, and uncheck all layers except for BasinLines.

5. Using the NHD stream pattern, use the "Edit" tool on the Editor toolbar (fig. 14) to select any digital lines already in the BasinLines feature class that are part of the new basin perimeter. Select features, then select "Edit," select "Copy" and then click "Edit," and select "Paste" to copy these features from BasinLines to NewBasinLines. In figure 22, the "Selection" tab of the Table of Contents window is shown with only one layer checked and selectable: BasinLines; two lines from this GIS dataset have been selected (shown in the cyan color).

6. If there are any lines in the WBD that are part of the new basin perimeter, make the WBD line feature class the only selectable layer and select and copy lines from the WBD line feature class into the NewBasinLines feature class. In the "Selection" tab of the Table of Contents window, make the WBD lines the only selectable layer (uncheck all layers except for the WBD lines), and copy and paste lines needed for the basin perimeter into NewBasinLines. Figure 23 shows the selection of a WBD line that is to be copied into the purple-colored NewBasinLines feature class.

7. In the Display tab window of the Table of Contents, check only the NewBasinLines feature class (and, optionally, also the medium-resolution NHD) to make it the only layer visible and zoom out to the full basin.

8. After copying lines into NewBasinLines, there are likely several lines that extend beyond the basin perimeter (fig. 24). These lines will need to be trimmed as described in the next few steps until only the lines that are part of the final basin perimeter remain.

9. In the Selection tab window of the Table of Contents, check only the NewBasinLines feature class to make it the only selectable layer, and set the snapping in the "Snapping Environment" window to "Vertex" of NewBasinLines. Zoom into the line intersection and select a line that has a section that extends beyond the basin boundary. On the Editor toolbar, select the "Split Tool" (fig. 14).

10. With the "Split Tool" active, click on the NewBasinLine vertex that separates the part of the line that belongs to the basin perimeter from the part that does not. After the line has been split at the vertex into two pieces, make only the NewBasinLines feature class selectable (use the Selection tab in the Table of Contents window and check the checkbox only for this feature class), and use the "Edit Tool" (fig. 14) to select the line that is external to basin boundary and delete it (fig. 25).

11. Continue to split and delete extraneous lines until there is a single, closed, basin perimeter as shown in figure 26.

12. To check that NewBasinLines is a single, closed loop, click on the "Validate Entire Topology" button on the Topology toolbar (fig. 19), and click "OK" to validate the entire topology. (The previously discussed "General Topological Editing Guidelines" section should be consulted at this stage.)

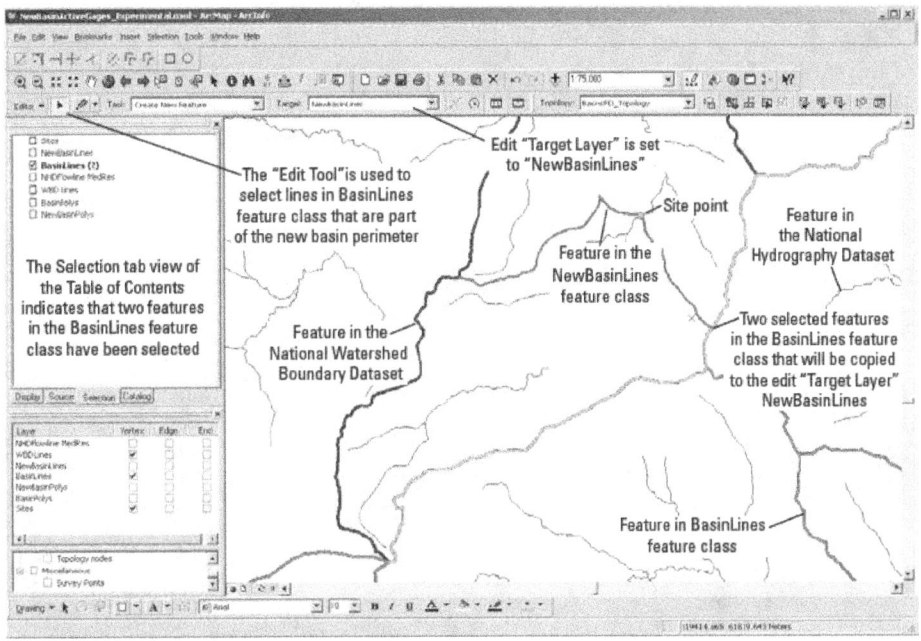

Base from U.S. Geological Survey digital data, 2012, 1:24,000
Albers equal-area conic projection
Standard parallels 37° 30' N and 40° 30' N
Central meridian 105° 00' W

Figure 22.   ArcMap view of copying existing lines from BasinLines to NewBasinLines.

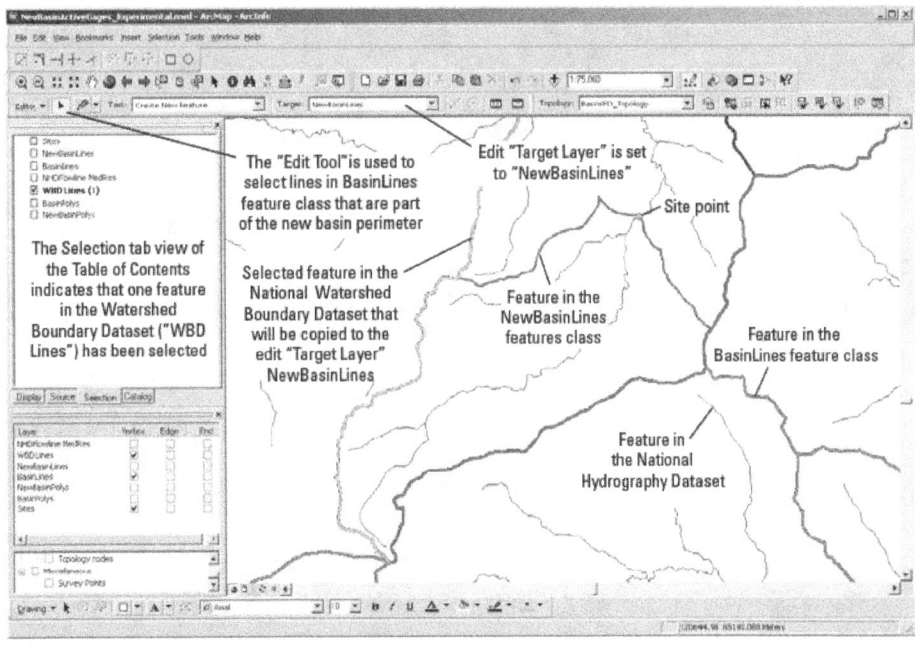

Base from U.S. Geological Survey digital data, 2012, 1:24,000
Albers equal-area conic projection
Standard parallels 37° 30' N and 40° 30' N
Central meridian 105° 00' W

Figure 23.   ArcMap view of copying a line from the National Watershed Boundary Dataset (WBD) to NewBasinLines.

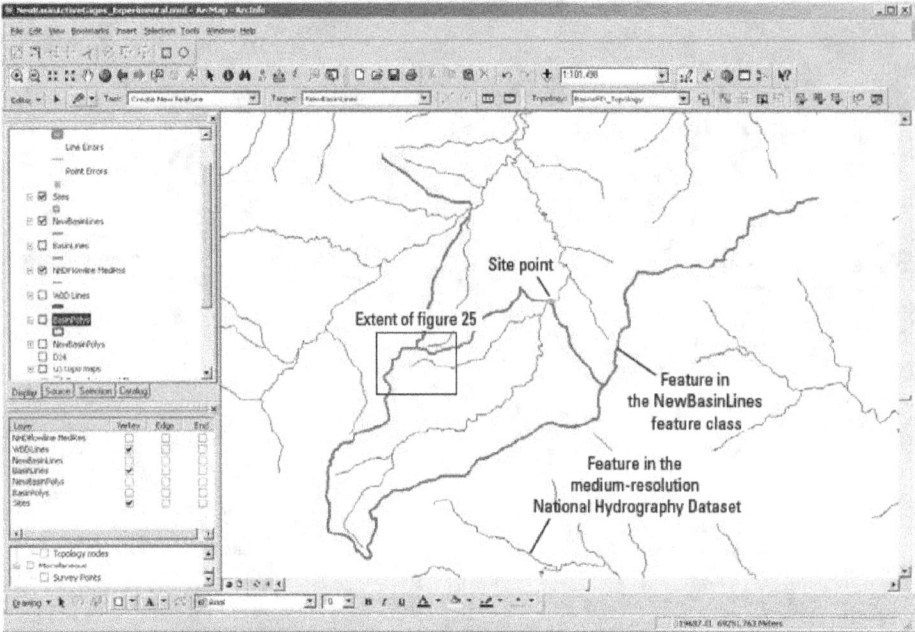

Base from U.S. Geological Survey digital data, 2012, 1:24,000
Albers equal-area conic projection
Standard parallels 37° 30' N and 40° 30' N
Central meridian 105° 00' W

Figure 24.   ArcMap view of the preliminary basin before trimming lines that are not part of the new basin boundary.

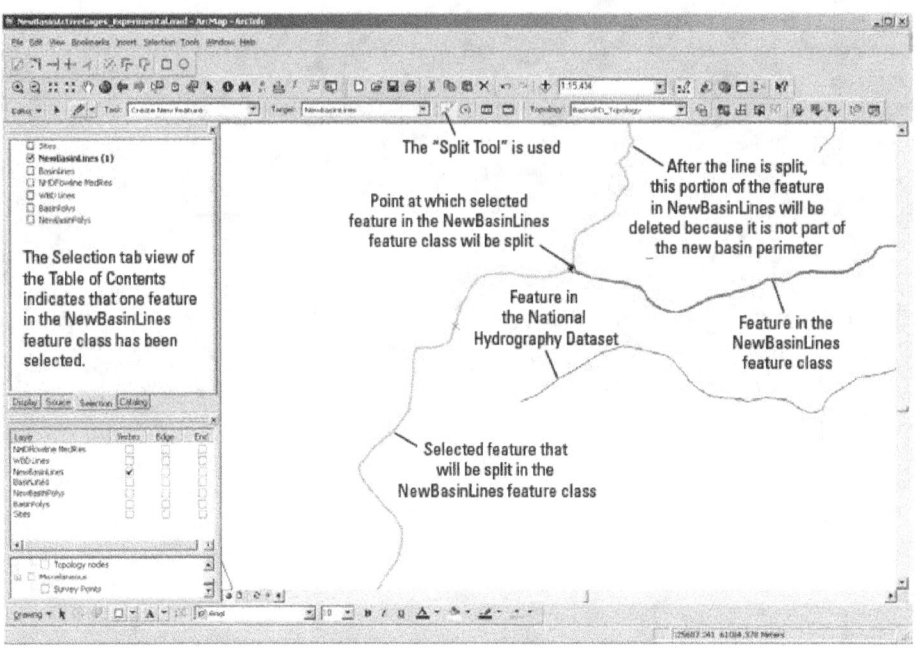

Base from U.S. Geological Survey digital data, 2012, 1:24,000
Albers equal-area conic projection
Standard parallels 37° 30' N and 40° 30' N
Central meridian 105° 00' W

Figure 25.   ArcMap view of a selected line to be split at a vertex using the "Split Tool."

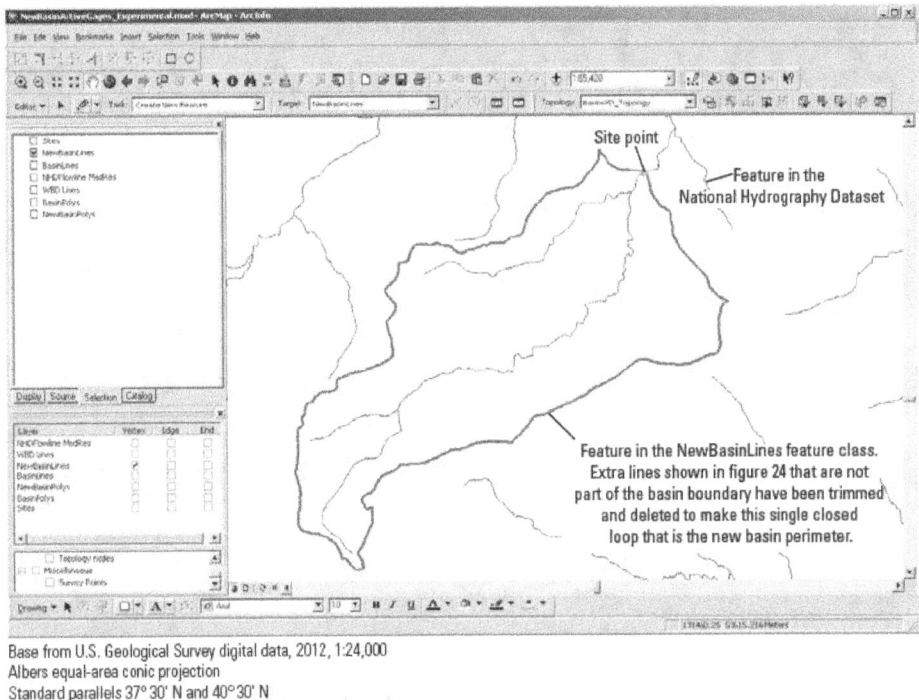

Base from U.S. Geological Survey digital data, 2012, 1:24,000
Albers equal-area conic projection
Standard parallels 37° 30' N and 40° 30' N
Central meridian 105° 00' W

Figure 26.   ArcMap view of the final basin perimeter after splitting and removing extraneous lines.

13. After validating the topology, check the errors by selecting the "Error Inspector" button on the Topology toolbar (fig. 19). Any errors will be depicted in the map view as pinkish line or point features if the BasinsFD_Topology is visible.

14. If there are multiple errors of all types at this stage, one or more lines from BasinLines may have been inadvertently "cut" instead of copied into NewBasinLines. Use the "Undo" button (fig. 17) to back out of the editing steps until the tool tip for this button shows an "Undo Cut" edit, and undo that edit. Redo the steps to copy and paste the lines into NewBasinLines, the steps to split and trim the lines to the closed basin perimeter, and validate the topology.

15. There should be multiple "NewBasinLines Must Be Covered By Feature Class Of BasinLines Polyline" errors, which will be dealt with in a later step.

16. The presence of any "NewBasinLines Must Not Have Dangles" errors indicates either that there are gaps in the new basin perimeter or that there are small line segments that still extend beyond the basin perimeter that need to be split and deleted from the perimeter. To repair this situation, view each dangle error by right clicking on one of the "NewBasinLines Must Not Have Dangles Point" errors, and selecting "Zoom." If a pair of dangle errors

surround a gap in the perimeter as in figure 27, select the line from BasinLines or from the WBD that was missed during the copying stage and copy and paste it into NewBasinLines. Paired dangle errors may indicate a line that is wholly outside the new basin perimeter; select the extraneous line and delete it.

17. Single dangle errors generally represent digitized lines that are overshoots from the basin perimeter. If the dangle is caused by a short extension of a line in the basin perimeter (where the extension is not part of the basin perimeter), select, split, and delete the extraneous line.

18. Validate the entire topology until there are no "NewBasinLines Must Not Have Dangles Point" errors, and until the only errors remaining are "NewBasinLines Must Be Covered By Feature Class Of BasinLines Polyline" errors (fig. 28). Such "errors" consist of new lines in the NewBasinLines feature class that will be added to BasinLines after first making the new BasinPoly feature.

19. Select all NewBasinLines and construct a new feature in the BasinPolys feature class. Make NewBasinLines the only selectable layer, set BasinPolys as the edit "Target Layer," and click on the "Construct Features" tool from the Topology toolbar (fig. 19).

20. In the "Construct Features" dialog window, ensure that the first radio button is selected to "Create new polygons from selected features."[22] Selecting this option will preserve nested basins in the BasinPolys feature class. Click "OK" to create the new feature.

---

[22]The radio button for the option to "Create new polygons (considering existing features in target layer)" should not be used. Doing so will flatten any nested basins in the geodatabase. If this accidentally occurs, try to use the "Undo" button to undo the "Construct Feature" or revert to a copy of the geodatabase.

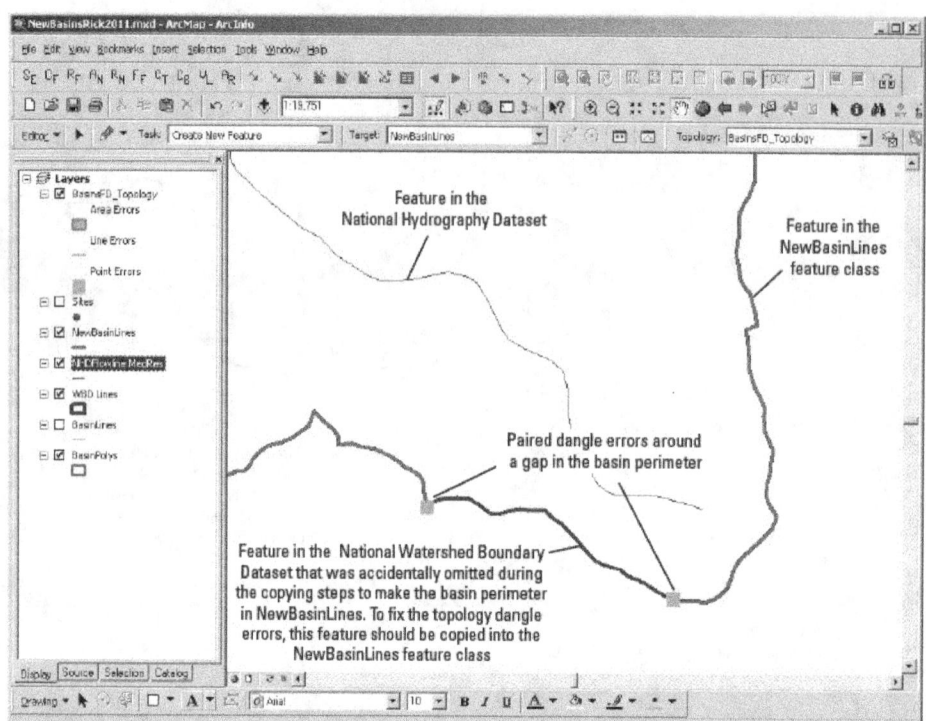

Base from U.S. Geological Survey digital data, 2012, 1:24,000
Albers equal-area conic projection
Standard parallels 37° 30' N and 40° 30' N
Central meridian 105° 00' W

Figure 27.   Map view of an example of paired dangle errors around a gap in the basin perimeter.

21. Clear the selected features in NewBasinLines by using the "Clear Selected Features" button in the Tools toolbar (fig. 18).

22. Click on the "Validate Entire Topology" button on the Topology toolbar (fig. 19). Open the "Error Inspector" window, and uncheck the "Visible Extent only" checkbox and click the "Search Now" button to view all errors. After validating the topology, be sure the topology, BasinsFD_Topology, is checked in the Table of Contents window to view the locations of topological errors.

23. The errors should include only "NewBasinLines Must Be Covered By Feature Class Of BasinLines Polyline" errors and also a single "BasinPolys Must Be Covered By BasinLines Polyline" error. Both errors indicate that BasinLine features for the new basin are missing: the first error indicates that there are features in NewBasinLines that are not yet in BasinLines; the second error indicates that a feature in the BasinPolys feature class lacks overlapping features in BasinLines. Both errors will be repaired in the next step.

24. Set the edit "Target Layer" to BasinLines, and highlight all the "NewBasinLines Must Be Covered By Feature Class Of BasinLines Polyline" errors in the "Error Inspector" window by right clicking on the top and bottom error while holding down the shift key. Right click and "Select Features," and copy and paste these features into BasinLines.

25. Validate the entire topology; there should be either no errors or only one type of remaining error. Correct any remaining "BasinLines Must Not Intersect or Touch Interior Point" errors (refer to the "Shape" column in the "Error Inspector" window to differentiate "Point" and "Polyline" errors for this topology rule). Point errors of this type represent intersections that lack a node.[23] Generally such errors are trivial, but these point errors should be scrutinized before fixing them because they represent the only diagnostic tool to prevent gaps and slivers between adjacent polygons, which occur where boundary lines for adjacent polygons weave around one another instead of exactly overlapping.

   a.   If there are only a few such point errors and they occur at obvious line intersections, particularly where a newly digitized line meets the WBD or meets a BasinLine, fix the error by right clicking and selecting "Split" to generate a new node at the line intersection.

   b.   If, on the other hand, there are many such errors spattered along a polygon boundary, one of the new lines added to the BasinLines feature class does not

---

[23]Such errors will occur at some intersections of the newly digitized arcs if they crossed a line in NewBasinLines that was copied from BasinLines. If the copied line was split in NewBasinLines so a piece of the copied arc could be deleted, a node is needed now in BasinLines at the point at which the arc was split.

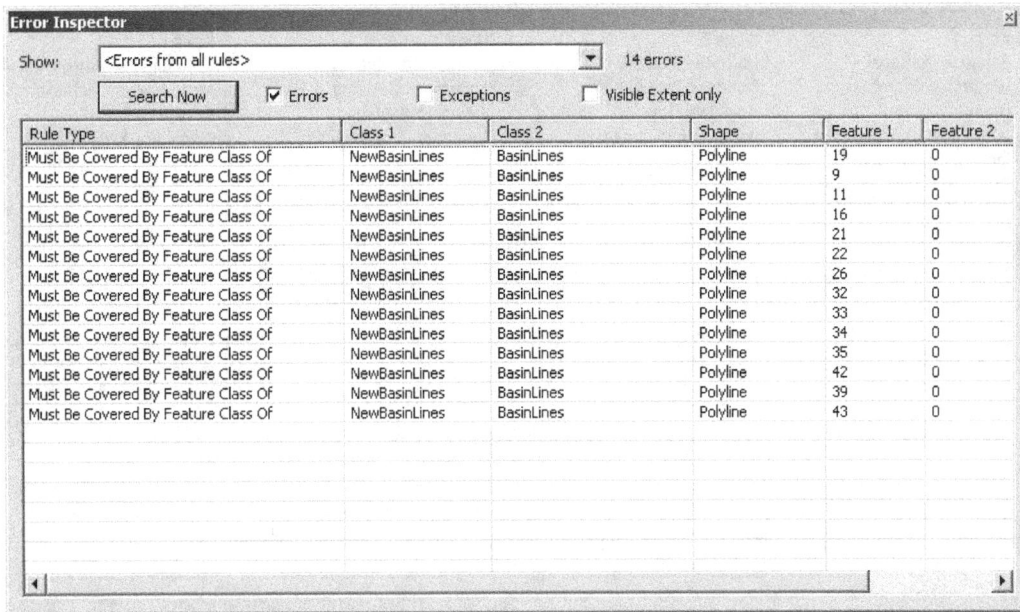

Figure 28. Error Inspector window showing "NewBasinLines Must Be Covered By Feature Class Of Basin Lines Polyline" errors.

exactly match existing lines in this feature class. This is a serious error. Either the new basin will need to be redelineated or one of the old basins in the geodatabase archive contains a line that has been moved and the old archived basin may need to be redelineated.

26. When the topology can be validated without errors, select and delete all features in NewBasinLines. This feature class should be empty before creating another basin perimeter.

27. Update the attributes and, if the entire topology can be validated without errors or if all dirty areas have been eliminated using the two other topology validation tools without unrepaired errors in the "Error Inspector" window (refer to the "General Topological Editing Guidelines" section above), save edits. The guidelines in the "Post-Delineation Steps" section above should be consulted at this stage.

Cut-Polygon Delineation Method

In essence, this delineation method involves combining WBD polygons into one polygon in NewBasinPoly, developing a "cutter line" (by digitizing and/or cobbling together digitized lines with existing BasinLines or WBD lines) in NewBasinLines that extends across this polygon at the basin drainage divide through the site point, and using the cutter line to cut out the desired basin polygon from the merged WBD polygon (fig. 29). Once mastered, this second delineation method is often a much quicker workflow and is the more useful workflow when constructing multiple nested basins. If delineating nested basins, first create the basin for the most downstream site point, and cut the basin polygon in NewBasinPoly successively upstream for additional site points on the same

main stem. The only possible danger of this method is that if the polygon that is being cut for nested basins is accidentally moved, it will produce a basin that is slightly displaced in the final archive. It is important to pay attention to the topology errors (especially the "BasinLines Must Not Intersect Or Touch Interior" "Point" and "Polyline" errors) in order to catch this error (and undo the steps that made it) should it happen.

1. The background information in the "Initial Editing Steps" and in the "General Digitizing Steps" sections should be consulted at this stage.

2. Add the NewBasinPoly feature class to the ArcMap session. In the edit "Target Layer" pull-down list of the Editor toolbar (fig. 14), select the empty NewBasinPoly feature. Ensure that the NewBasinLines feature class also has been added.

3. Click on the "Selection" tab in the Table of Contents window to make the WBD polygons the only selectable layer.

4. Select all WBD polygons that contribute water to the site point, including the most downstream WBD polygon that includes the site point. The NHD stream network can be used to guide selection of these polygons. The goal is to combine these multiple polygons into one feature in the NewBasinPoly feature class. This combined feature will be the digital "cookie" that will be cut to the final dimensions of the new polygonal basin area. Make NewBasinPoly the edit "Target Layer."

5. After multiple WBD polygon selections are made, combine the selected polygons into a single feature in NewBasinPoly by selecting the "Editor" on the Editor toolbar (fig. 14), and selecting "Union" from the pull-down menu. For large

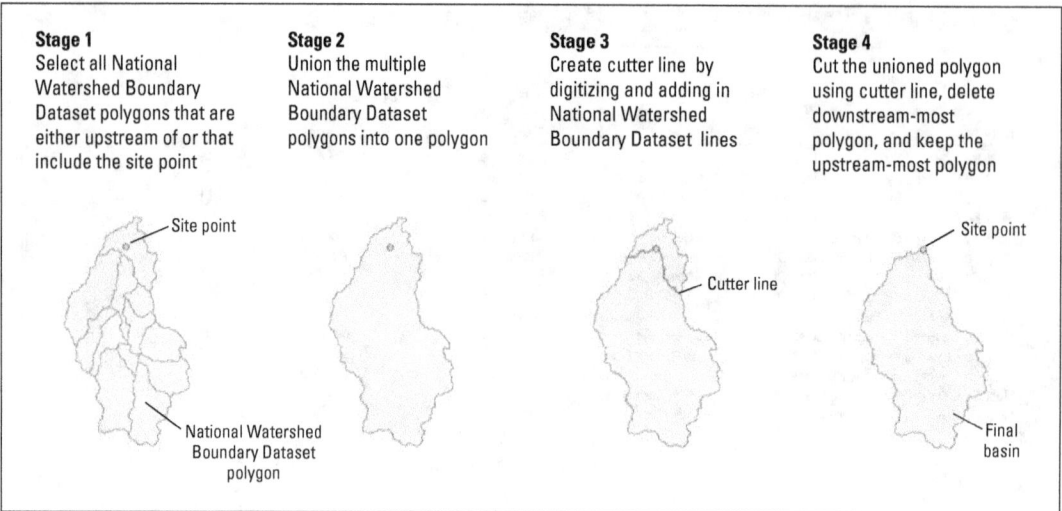

Stage 1
Select all National Watershed Boundary Dataset polygons that are either upstream of or that include the site point

Stage 2
Union the multiple National Watershed Boundary Dataset polygons into one polygon

Stage 3
Create cutter line by digitizing and adding in National Watershed Boundary Dataset lines

Stage 4
Cut the unioned polygon using cutter line, delete downstream-most polygon, and keep the upstream-most polygon

Site point

National Watershed Boundary Dataset polygon

Cutter line

Site point

Final basin

Figure 29.   Schematic of the four stages of basin delineation using the cut-polygon method.

basins, multiple, combined polygons in NewBasinPoly created from several phases of using the "Union" function can be merged into one polygon feature by selecting all the features in NewBasinPoly, selecting "Editor" on the Editor toolbar (fig. 14), and selecting "Merge" from the pull-down menu. In figure 30, one WBD polygon needed for a small basin has been copied into the NewBasinPoly feature class.

6. After making the single NewBasinPoly polygon, a "cutter line" is needed to cut the NewBasinPoly into two pieces, one of which will be kept as the final polygonal drain-age area. The cutter line includes the site point and is the downstream-most basin boundary that stretches across the single polygon in NewBasinPoly. In this delineation method, the NewBasinLines feature class is used to store the cutter line, which consists of lines digitized from the site point plus any additional BasinLines or WBD lines needed to make the cutter line extend all the way across the single polygon feature in NewBasinPoly.

7. The digitizing steps and guidelines in the "General Digitizing Steps" section should be consulted at this stage. Two lines need to be digitized from the site point to the nearest BasinLines feature, WBD line, or toward the boundary of the polygon in NewBasinPoly, whichever is the closest feature. As each line is digitized from the site point, one of two possible situations will be encountered.

    a. If the closest line to the basin-site point is the bound-ary of the polygon feature in NewBasinPoly (and not an intervening BasinLine feature or a WBD line), snap the last digitized vertex either to an existing ver-tex in NewBasinPoly (in the "Snapping Environment" window, check "Vertex" for NewBasinPoly) or to a line segment between vertices, called an "edge" (in the "Snapping Environment" window, check "Edge" for NewBasinPoly). Immediately after this last point is digitized, right click, and select "Finish Sketch" to stop digitizing. Figure 31 shows a situation where the

closest feature to the site point is the boundary of the polygon feature in NewBasinPoly. A line has been digitized so that the last vertex of the digitized line has been allowed to snap to the boundary of the feature in NewBasinPoly.

    b. The second possible situation when digitizing the cutter line is that there will be one or more inter-vening BasinLines features or WBD lines along the basin divide between the digitized line and the polygon boundary in NewBasinPoly. If this is the case, snap the last vertex on the digitized line either to the "Vertex" or to an "Edge" of the nearest feature in the BasinLines or WBD line feature class (using the "Snapping Environment" window checkboxes). Immediately after this last vertex is digitized, right click, and select "Finish Sketch" to finish digitiz-ing. After digitizing this segment from the site point, make either the BasinLines or WBD line feature class selectable, and copy and paste any additional lines needed to make the cutter line (the downstream end of the basin) extend from the site point in both directions and terminate it at the NewBasinPoly boundary.

8. If one or more NewBasinLines were copied from BasinLines or from the WBD, these features may need to be trimmed at the point where they are intersected by a digitized line to include only the section that is part of the new basin perimeter. To trim the copied line, make the NewBasinLines feature class the only selectable layer and set the options in the "Snapping Environment" window to snap to an "End" in NewBasinLines (the end of the digitized line in NewBasinLines where it meets a copied feature). Select the copied line in NewBasinLines, select the "Split Tool" (fig. 14) on the Editor toolbar, and click on the vertex at the end of the digitized line. After splitting the line, select the line that is extraneous to the new basin perimeter and delete it using the "Delete" button (fig. 17).

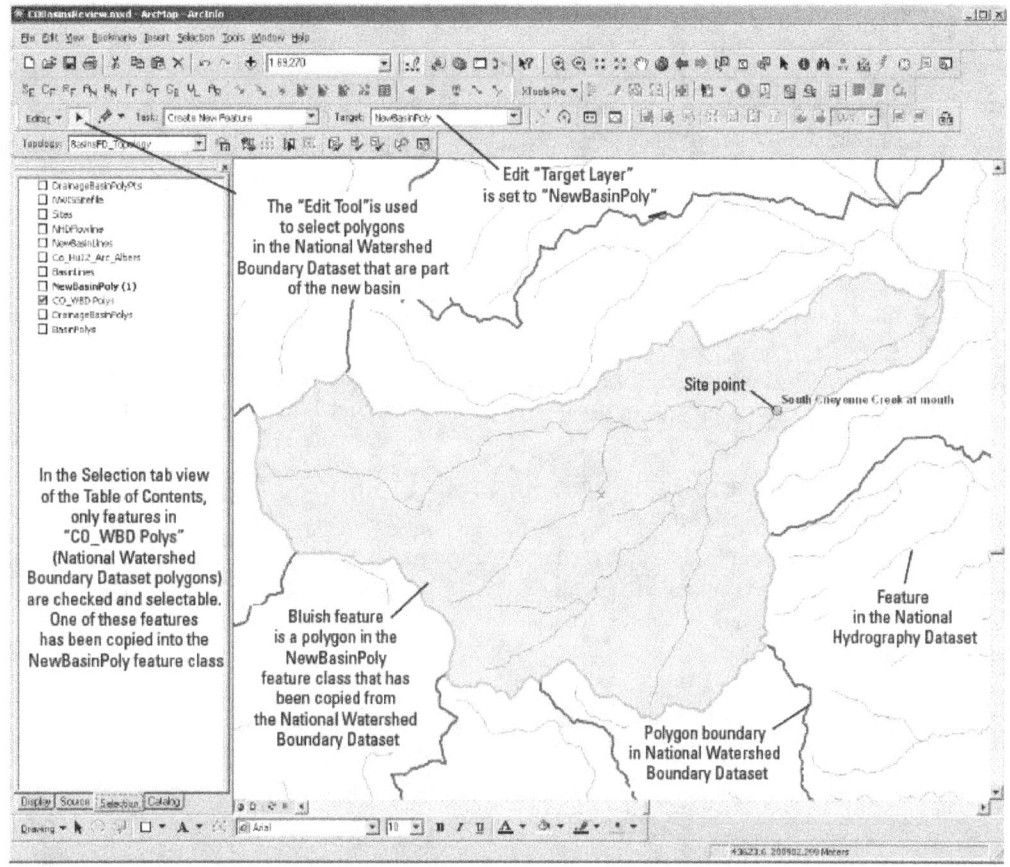

Base from U.S. Geological Survey digital data, 2012, 1:24,000
Albers equal-area conic projection
Standard parallels 37° 30′ N and 40° 30′ N
Central meridian 105° 00′ W

Figure 30.   Single National Watersheds Boundary Dataset (WBD) polygon added to the NewBasinPoly feature class.

9. The next few steps are used to check for missing segments in the NewBasinLine cutter line using the "Validate Entire Topology" button (fig. 19) on the Topology toolbar. After validating the topology, be sure the topology, BasinsFD_Topology, is checked in the Table of Contents window to view the locations of topological errors. The topology errors will be symbolized on the map as either pink-colored line, point, or polygon features. The guidelines in the "General Topological Editing Guidelines" section should be consulted at this stage.

10. Click on the "Error Inspector" button (fig. 19) to view the "Error Inspector" window. Uncheck the "Visible Extent only" checkbox in the "Error Inspector" dialog window, and select the "Search Now" button to view all errors.

11. After validating the topology, there will be several "NewBasinLines Must Be Covered By Feature Class Of Basin Lines Polyline" errors, which can be ignored at this stage. There will also be two "NewBasinLines Must Not Have Dangles Point" errors, which should occur

at each end of the cutter line (on the boundary of the polygon in NewBasinPoly). If, however, there are more than these two dangle errors, there may be gaps in the NewBasinLines feature class, and these need to be filled by copying missing short lines from the BasinLines or from the WBD lines feature class. After copying these lines, validate the entire extent of the topology, and there should be only two dangle errors at the ends of the overall feature in NewBasinLines plus the "NewBasinLines Must Be Covered By Feature Class Of Basin Lines Polyline" errors.

12. At this stage, there will be several features in NewBasinLines (at least two lines digitized from the site point and possibly also several copied from the BasinLines feature class or from the WBD). Collectively, these lines should form the downstream-most perimeter of the new basin and should stretch across the polygon feature in NewBasinPoly. The cutter line must be a single feature to cut the polygon, therefore multiple lines need to be merged. To combine all the lines in NewBasinLines into a single cutter line, select all the features in NewBasinLines,

and from the Editor toolbar (fig. 14) select "Editor" and then "Merge" from the pull-down menu options. In the "Merge" dialog window, select any feature in the list, and select "OK" to merge all these features into one cutter line that extends across the full width of the polygon feature in NewBasinPoly (fig. 32). Clear the selection using the "Clear Selected Features" button in the Tools toolbar (fig. 18).

13. The next few steps guide the process of cutting the polygon using the cutter line. In the Table of Contents window, it is helpful to turn off the visibility of any layers (especially of any polygon layers) except for NewBasinLines and NewBasinPoly (and optionally, the DRG maps). Right click on NewBasinLines and zoom to the extent of the cutter line.

14. In the "Selection" tab in the ArcMap Table of Contents window, make only NewBasinPoly selectable.

15. Change the "Edit Task" to "Cut Polygon Features" (the edit "Target Layer" can be set to anything; because the final polygon will be added to BasinPolys, this can be set as the "TargetLayer").

16. Select NewBasinPoly, then using the "Sketch Tool" (fig. 14), right click on the cutter line in NewBasinLines. Select "'Replace Sketch" among the options, and the cutter-line vertices will appear (fig. 33).

17. Right-click on one side of the NewBasinLines cutter line and inside the selected polygon in NewBasinPoly and select "Finish Sketch." The polygon should be cut into two polygons.

18. If the cut-polygon operation fails, the NewBasinLines cutter line contains at least one gap or its endpoints do not quit extend to the edge of the polygon to be cut. Add the "Advanced Editing" toolbar (fig. 34) to ArcMap (click View, click Toolbars, and put a check mark by the "Advanced Editing" toolbar in the list of available toolbars). Select the merged feature in the NewBasinLines feature class, and click on the "Explode Multi-part Feature" tool in the toolbar to separate the cutter line into discrete components. Any gaps in the NewBasinLines cutter line will be found in between the multiple features in NewBasinLines created after using this tool. Copy and paste the missing line features from the BasinLines or the

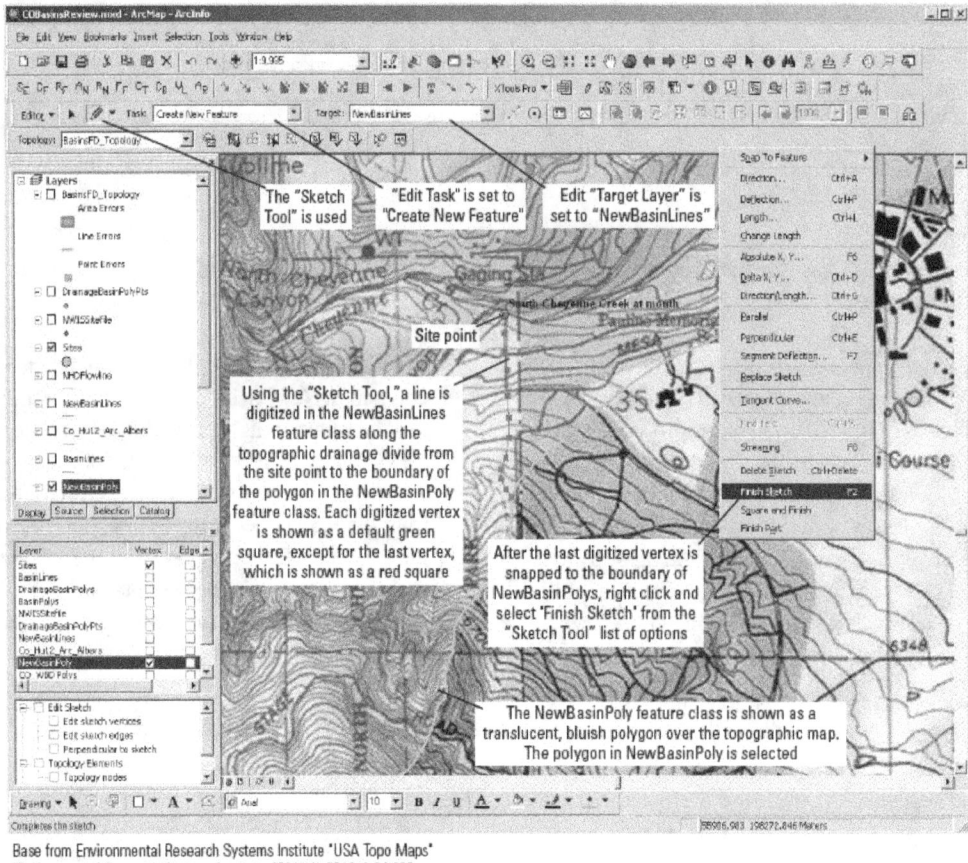

Base from Environmental Research Systems Institute "USA Topo Maps" (Environmental Research Systems Institute, [2012b]), 2012, 1:24,000
Albers equal-area conic projection
Standard parallels 37° 30' N and 40° 0' N
Central meridian 105° 00' W

Figure 31.   ArcMap view of Line digitized in NewBasinLines from the site to the boundary of the polygon in NewBasinPoly.

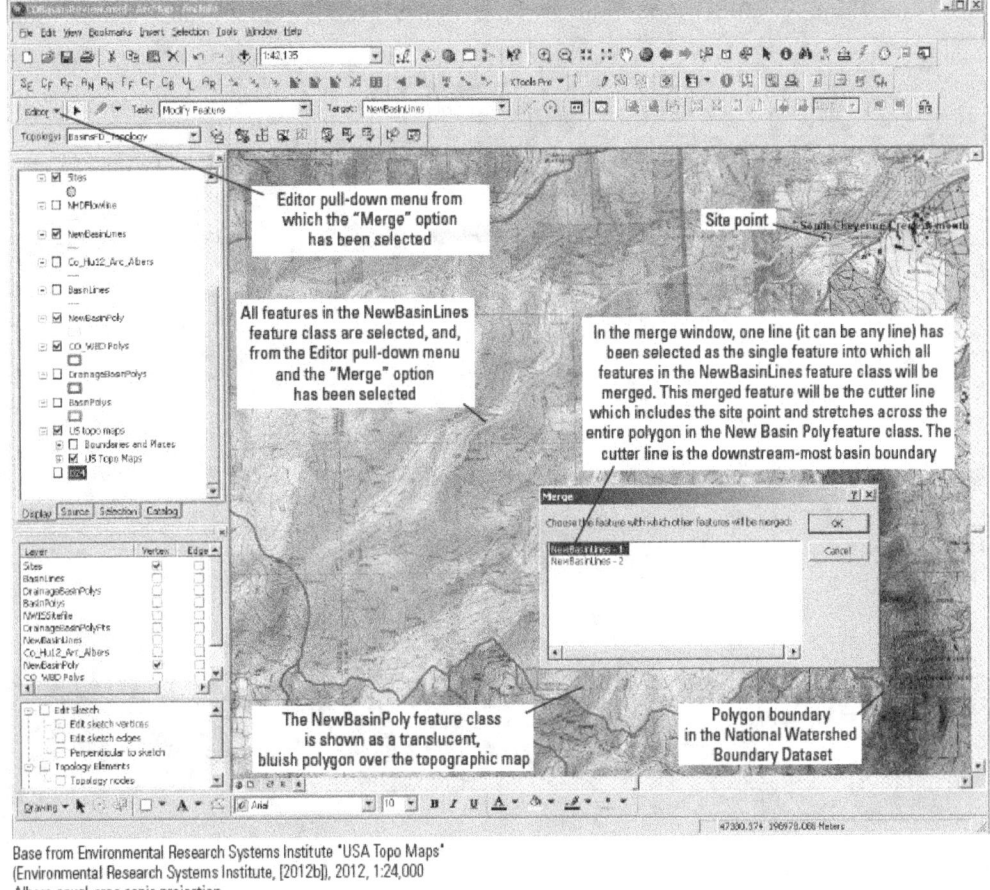

Base from Environmental Research Systems Institute "USA Topo Maps"
(Environmental Research Systems Institute, [2012b]), 2012, 1:24,000
Albers equal-area conic projection
Standard parallels 37° 30' N and 40° 30' N
Central meridian 105° 00' W

Figure 32.   ArcMap view of using the "Merge" window to create the single cutter line in NewBasinLines.

WBD into the gaps in the cutter line, and then select all features in the NewBasinLines feature class and re-merge them. If, after exploding the feature, there is only a single feature in NewBasinLines, extend the cutter line slightly past the polygon boundary. Afterwards, select the feature in NewBasinPoly to be cut, repeat the cutting steps above, and the polygon should be cut into two pieces.

19. Make NewBasinPoly the only selectable layer, and, of the two polygons in NewBasinPoly, select the polygon downstream from the site point and cutter line and delete it.

20. Make BasinPolys the edit "Target Layer," select the remaining polygon in NewBasinPoly that is upstream from the site point, copy and paste it into BasinPolys, and clear all selections using the "Clear Selected Features" button in the Tools toolbar (fig. 18).

21. If the cutter line feature in NewBasinLines includes non-digitized lines copied from BasinLines and/or the WBD, split and delete these features from the cutter line. To do this for each end of the digitized line, make NewBasinLines the only selectable layer, change the snap features in the" Snapping Environment" window to

snap only to the "Vertex" of NewBasinLines. Select the NewBasinLine, and use the "Split Tool" (fig. 14) to split the digitized lines at the vertices where they join lines copied into NewBasinLines. After splitting the cutter line at each end of the digitized section, delete any non-digitized portions of NewBasinLine. If the cutter line was extended past the polygon boundary to cut it, trim each end of the cutter line back to the polygon boundary before the next step.

22. Make BasinLines the edit "Target Layer," make NewBasinLines the only selectable layer, select the remaining (new, digitized) features in NewBasinLines, and copy and paste them into BasinLines. Clear the selection.

23. Select the features in NewBasinLines and delete them.

24. Make the BasinFD_Topology layer visible to view topology errors. Validate the entire topology using the "Validate Entire Topology" button on the Topology toolbar, and open the "Error Inspector" window to view the errors. If possible, view the "Error Inspector" window in a second monitor.

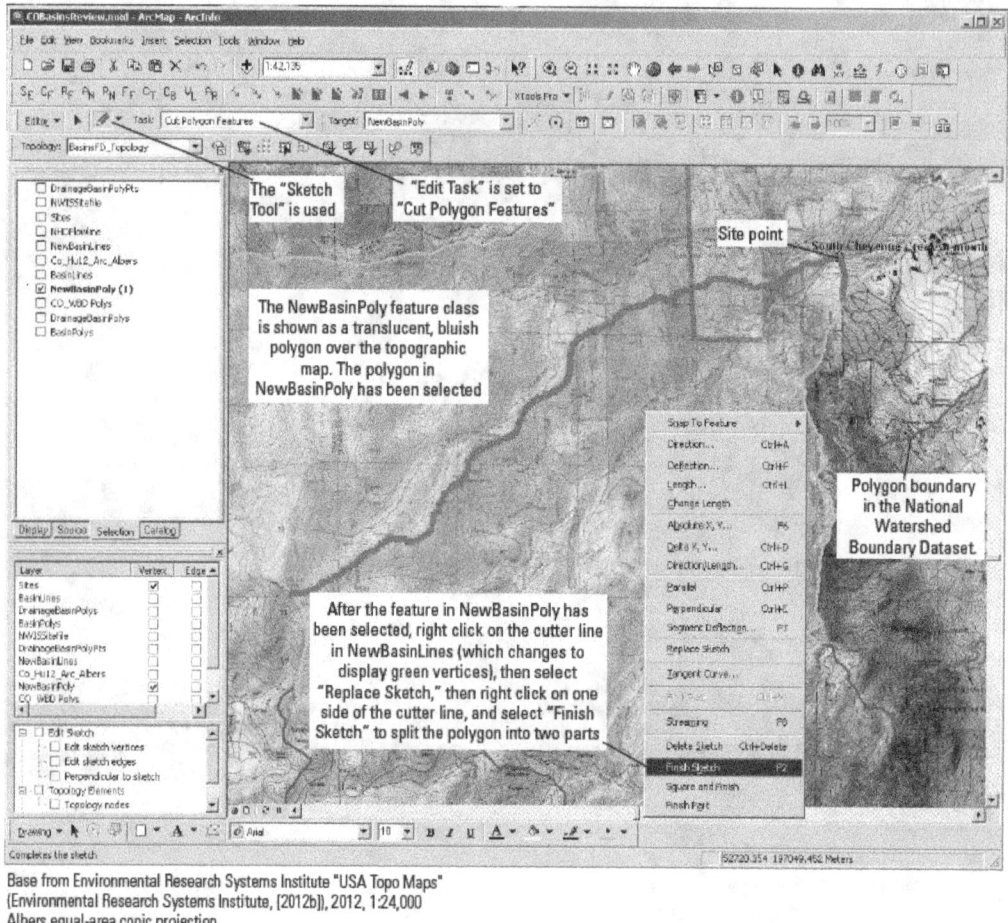

Base from Environmental Research Systems Institute "USA Topo Maps"
(Environmental Research Systems Institute, [2012b]), 2012, 1:24,000
Albers equal-area conic projection
Standard parallels 37° 30' N and 40° 30' N
Central meridian 105° 00' W

Figure 33. ArcMap view of a portion of the process for cutting a basin polygon.

25. At this stage there may be several "BasinPolys Boundary Must Be Covered By BasinLines Polyline" errors associated with one or more adjacent "BasinLines Must Not Have Dangles Point" errors. Generally, these collectively represent segments of the newly added BasinPolys feature where there are no overlying features in BasinLines. Such errors generally also represent segments of the polygon boundary where any features copied from the WBD need to be added to the BasinLines archive. Each of these polyline errors should overlie a WBD feature, and at least one end of the polyline error should terminate at a line intersection in the WBD. Most importantly, there must not already be a BasinLines feature at any of these locations. To check this, make BasinLines and the BasinsFD_Topology layers both visible, and select each of these polyline errors, one at a time, in the "Error Inspector" window (the selected topology polyline error will be depicted as a black line in the map view). Turn off the BasinsFD_Topology layer to see if there is a BasinLines feature underneath the polyline error. If there is a BasinLines feature underneath part or all of the

black polyline error in BasinsFD_Topology, redelineate the basin. Often in this situation there will be many such polyline errors spattered around the entire basin boundary (and also many associated "BasinLines Must Not Intersect Or Touch Interior Point" errors) indicating that the new basin boundary is different from other lines and polygons in the archive, and in this situation it is best to start over or use the basin perimeter delineation method. If there is no BasinLines feature under the black-colored polyline error, fix the "Basin Poly Boundary Must Be Covered By BasinLines Polyline" error by right clicking on the error and selecting "Create Feature" to create a new feature in

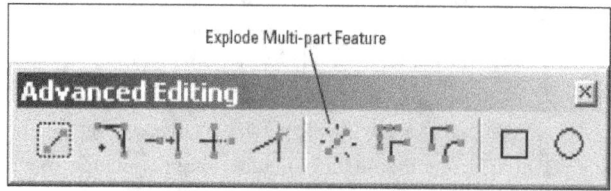

Figure 34. The ArcMap Advanced Editing toolbar.

BasinLines. After checking and fixing all these polyline errors, validate the entire topology (without repairing the other errors), and click the "Search Now" button in the "Error Inspector" dialog window.

26. After these fixes have been made and topology has been validated, generally the only remaining errors are "BasinLines Must Not Intersect Or Touch Interior Point" errors. Like the previous error, such errors can be either trivial or important. This point error indicates that a linear intersection has no node, and it usually appears where a line digitized from the site point meets an arc that existed in BasinLines before the new lines were added (click on the error in the "Error Inspector" window to view where the error is: the selected error will appear as a black square; alternatively, use the "Fix Topology Error Tool" in the Topology toolbar (fig. 19) to select individual errors on the map). If such errors are few and occur only at line intersections, especially where new lines intersect a line feature that was already in BasinLines before the new basin was delineated, they are fixed by right clicking on the error in the "Error Inspector" window, selecting "Split," and then revalidating the entire topology. However, numerous topological errors of this type spattered around the new basin boundary combined with other errors of various types indicate that the new basin boundary is shifted away from features already in the BasinLines and BasinPolys. Such errors indicate places where the newly added lines and old lines cross and weave around one another. They are fixed by undoing previous editing steps to add the polygon and cutter line and then redelineating the basin. Hopefully, the shift was simply caused by slightly moving a polygon when adding it and will not re-occur if the basin is redelineated. If basin boundaries have been moved in the archive from the WBD boundaries, it may be necessary to redelineate an older, archived basin that is offset from the actual WBD hydrologic unit boundaries.

27. A less common error is the "BasinLines Must Not Intersect Or Touch Interior Polyline" error. Generally these are two lines that exactly overlap, and to fix the error, right click on the error, select "Subtract." In the subtraction window, select either line listed in the dialog window and click "OK," and then validate the topology. However, numerous swaths of these polyline errors combined with multiple "BasinLines Must Not Intersect Or Touch Interior Point" errors indicate areas of overlapping, braided lines. Such errors are fixed by undoing or deleting both the incoming polygon and line, deleting the features in NewBasinLines and NewBasinPoly, and starting over.

28. Validate the entire topology, and check again for errors in the "Error Inspector" window. Only after all errors are fixed and the entire topology has been validated (or all dirty areas are eliminated using the other topology validation tools; refer to the "General Topological Editing

Guidelines" section) is it safe to save without the potential of having unintentionally created the possibility of having to do more editing later to fix unknown topological errors.

29. Save edits only after validating the entire topology or dirty areas and reviewing errors in the "Error Inspector" window shows no remaining errors.

30. After the topology errors are fixed, make NewBasinPoly the only selectable layer, select the polygon feature in NewBasinPoly, and delete it. The NewBasinPoly and NewBasinLines feature classes can be then reused to delineate the next new basin.

31. Update attributes in the BasinPolys and BasinLines feature classes. The guidelines in the "Post-Delineation Steps" section above should be consulted at this stage. Because the line feature in NewBasinLines was combined, the new line feature may need to be selected and split to update the source correctly. If the feature is split, validate topology. If the topology can be validated without errors, save edits.

Steps to Finalize the Delineated Basin

At this stage of the basin delineation process, the goal is to review the polygonal basin area and the derived numeric drainage area values, determine if they are reasonable, and enter the GIS-derived numeric drainage area into the NWIS database. Several checks are performed including comparison of the new polygonal basin area to the stream pattern in the NHD and also comparing the new numeric drainage area to published NWIS areas.

For new basins, the polygonal basin area is first compared to the medium- and high-resolution NHD. The numeric drainage area and the NHD are made visible in the ArcMap Table of Contents window, and the stream pattern in the NHD is checked to be sure that streams that drain to the site point are all inside the new polygonal basin area. The basin boundary should not cross any tributaries of the stream on which the site point is located; if it does do so, these have been omitted as the WBD features were copied and they need to be added to the polygonal basin area.

Newly digitized lines and numeric drainage areas are also checked by the USGS Water Science Center Data Section who are the most familiar with sites and numeric drainage areas for a given state. If the newly delineated basin has a published USGS NWIS numeric drainage area (U.S. Geological Survey, 2012a), the GIS-derived numeric drainage area is compared to the published NWIS numeric drainage area. Generally, if the percent difference exceeds a threshold of 3 percent, the GIS basin boundary is compared to the basin boundary traced on hard-copy topographic maps, and the reasons for any discrepancies are determined. For such basins, digitized lines and even WBD lines are also re-checked against digital topography. (Often, the USGS CWSC inspects the basin boundaries if the raw difference between the NWIS and GIS areas exceeds 30 mi².) If the new GIS basin is for a site that does not yet have an NWIS area, the numeric GIS area is compared to known NWIS areas for basins upstream or downstream from the new GIS basin.

If the new numeric drainage area seems reasonable when compared to known NWIS areas and the newly digitized lines are considered to be correct, the polygon feature class attribute table field "Approved" is updated to a "Y" value to indicate that the numeric GIS-derived numeric drainage area is ready to be entered into NWIS. The USGS CWSC adds another optional "NWISChanged" flag field, having a "YesNo" domain with a default value of "N," to indicate whether the new, approved numeric drainage area in the "AreaRounded" field has been entered into NWIS. There are instances for which this field is left "N" for approved polygonal basin areas, such is if the site is managed by another water science center or if the basin was delineated for a site located on an artificial feature. If, after comparison with published NWIS areas and line checking, the new polygonal basin area is found to be in error, there are corrective measures that can be taken, which are described below.

Editing Problem Areas

Several common editing needs are specifically addressed in this last section. These edits pertain to changing features that are already in the archive and include changing a section of a basin boundary, moving the end of a line to a new location, and removing a polygonal area (cutting polygons) that should not be part of the archived basin polygon.

Changing a Section of a Basin Boundary

After basin delineation, some adjustments may need to be made to sections of a basin boundary in order to correct errors or to better conform to topographic contours. Such changes are generally not made to basin boundaries copied from the WBD but to boundaries that were digitized. Because such lines can form the perimeter of more than one basin, such changes are made using topology editing tools to preserve the spatial relations that should exist among all line and polygon features that share the basin boundary.

The "Topology Edit Tool" (fig. 19) in the Topology toolbar is generally used to make such edits. The selection made using this tool is a different selection than the regular ArcMap selection made using the "Edit Tool" (fig. 14) or the "Select Features" tool on the ArcMap Tools Toolbar (fig. 18).

A topological selection made using the "Topology Edit Tool" (fig. 19) is by default displayed in a magenta color. Unfortunately, a single feature, simultaneously, can be selected as both the regular ArcMap selection (by ESRI default, displayed with a cyan color) and also as a topological selection,[24] and the regular selection can 'hide' under the magenta-colored selection. The ability to select a feature as a regular ArcMap feature selection allows individual features to be moved and edited independently from features to which they are topologically related, and such editing is sometimes useful. But more commonly the act of selecting and moving individual features,

using the regular ArcMap selection, will separate them from features that they should overlie topologically and trigger a group of topological errors. If a feature has been selected as both types of selections and the "Edit Tool" (fig. 14) is active and the feature is moved with the mouse cursor, the feature will be completely detached from other line and polygon features in the archive. Therefore, it is a good idea to make it impossible to perform the basic ArcMap feature selection before making topological selections and edits using the "Topology Edit Tool." To do this, select the Selection tab in the ArcMap Table of Contents window and uncheck all of the layers to make sure that no layer is selectable. Then, if features need to be edited, use the "Topology Edit Tool," and features can be selected topologically only and edited as a unit.

Even if steps are taken to make sure that the more basic ArcMap feature selection cannot occur, it is recommended that you draw a selection box to make selections. Selecting a line using the "Topology Edit Tool" by right clicking once on the line will make a topological selection, but also it can shift the line, often imperceptibly. If the shift is not detected quickly, a suite of topological errors are created that are costly in terms of time to repair once they are saved (fig. 35). Accidentally moving an entire, topologically selected line in BasinLines separates it completely from other lines in the BasinLines feature class. The polygon boundary stretches at both ends of the line to account for the offset, but in the stretched area, no lines in BasinLines overlie the polygon boundary. If the edits are not yet saved, the "Undo" button can be used to back out of the mistake of moving the line.

When editing archived lines and polygons, the "Edit Tasks" generally are limited to the "Modify Edge" and the "Reshape Edge" tasks. These tasks do not function properly unless the "Topology Edit Tool" (fig. 19) has been used to select a line, called a topological "edge." Settings for selectable layers and for the edit "Target Layer" are irrelevant if "Topology Edit Tool" and the "Modify Edge" or "Reshape Edge" "Edit Tasks" are used in conjunction with the persistent topology rules.

To change the location of a line feature, select the line by drawing a selection box around part of the line using the "Topology Edit Tool." A message will appear, "Building topology cache on visible extent" that indicates that topology rules are being applied "on-the-fly" for the features in the map extent. If edits are needed for only a short length, use the "Modify Edge" "Edit Task" to move individual vertices. When the "Edit Task" is set to "Modify Edge" and a line is selected with the "Topology Edit Tool," vertices on the line should appear as the default ESRI vertex symbol: green squares. Position the mouse cursor over a green-square vertex until

---

[24]The cyan selection color is overprinted by the magenta topological selection color and 'hides' under the topology selection. A good way to know if a cyan-type of selection is active is to add a clear selection button to the Standard toolbar. Right click on the map view, select Customize (at the bottom of the list of toolbars), choose the Commands tab, and scroll down to Selection, click on Selection, click on Clear Selected Features and drag the tool to the Standard Toolbar, and Close the Customize dialog. If this tool is grayed out, there is no selection.

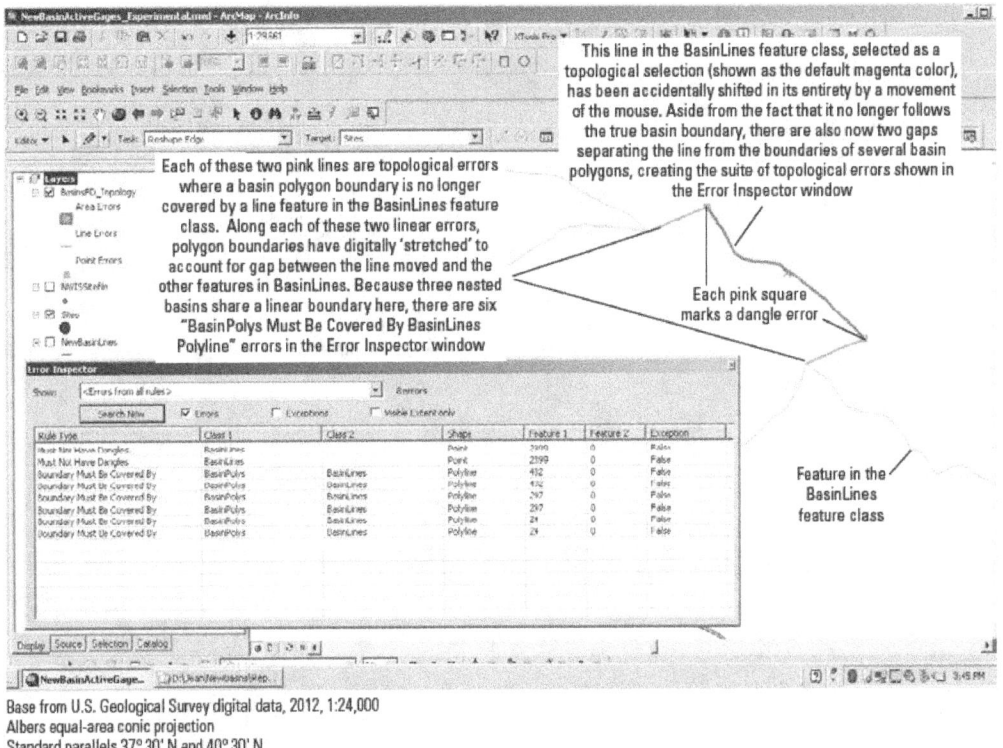

Base from U.S. Geological Survey digital data, 2012, 1:24,000
Albers equal-area conic projection
Standard parallels 37° 30' N and 40° 30' N
Central meridian 105° 00' W

Figure 35.   ArcMap view of resulting topology errors when a line selected as a topological selection is accidentally moved.

the cursor symbol changes to a diamond (or compass-like) symbol, and then drag the vertex to the preferred, new location (fig. 36). If the mouse cursor has not changed to the diamond symbol, the entire line will move. In that case use the "Undo" button and try again to select the vertex to be relocated. Move vertices until the line is positioned correctly. Right click along the line and select "Insert Vertex" to add a vertex; right click on a vertex and select "Delete Vertex" to remove a vertex. After editing, click away from the line while the "Topology Edit Tool" is active to unselect the line. Also, after editing, validate the entire topology using the "Validate Entire Topology" button on the Topology toolbar (fig. 19) to make sure there are still no errors. If validation of the topology takes longer than 10 or 15 minutes, refer to the "General Topological Editing Guidelines" section, and validate the topologically dirty area in sections. There should be no errors or possibly only a few "BasinLines Must Not Intersect Or Touch Interior Point" errors, which can be repaired by right clicking on each error and clicking on the option to "Split."

If changes to a line need to be made over a relatively long distance, use the "Reshape Edge" "Edit Task." For the reshaped line to be "accepted," the reshaped line must cross the original line twice: once near the place where the edits to the line begin and once just before the edits end. Click the "Topology Edit Tool," wait for the topology cache to be built for the map extent, select the line, select the "Sketch Tool" (fig. 14), click the mouse on one end of the line to be changed, cross the line with the mouse, and click to add the second vertex on the other

side of the line. Digitize the line with the "Sketch Tool," and when finished, cross the original line again, click to add a last vertex, right click, and select "Finish Sketch" (fig. 37). To start over while digitizing a line, right click and select "Delete Sketch," and start again. After the new line has been digitized, click away from the line using the "Topology Edit Tool" to clear the topological selection. Validate the entire topology, and there should be no errors. If validation of the topology takes longer than 10 or 15 minutes, refer to the "General Topological Editing Guidelines" section, and validate the topologically dirty area in sections. There should be no errors or possibly only a few "BasinLines Must Not Intersect Or Touch Interior Point" errors, which can be repaired by right clicking on each error and clicking on the option to "Split."

Occasionally, a node (a line intersection point) needs to be moved. Such fixes are generally made only for digitized lines. To move a node, click on the "Topology Edit Tool," and click on the map view to build the topology cache. Press the keyboard key "n," and draw a small selection box around the node. Once the node is selected (it will appear as a magenta-colored point), move the cursor over the node and wait for it to change to the diamond-shaped cursor symbol, and only then drag the node to the new location (fig. 38). Click away from the node with the "Topology Edit Tool" active to clear the selection. After editing, validate the entire topology, and there should be no errors. If validation of the topology takes longer than 10 or 15 minutes, refer to the "General Topological Editing Guidelines" section, and validate the topologically dirty area in sections. There

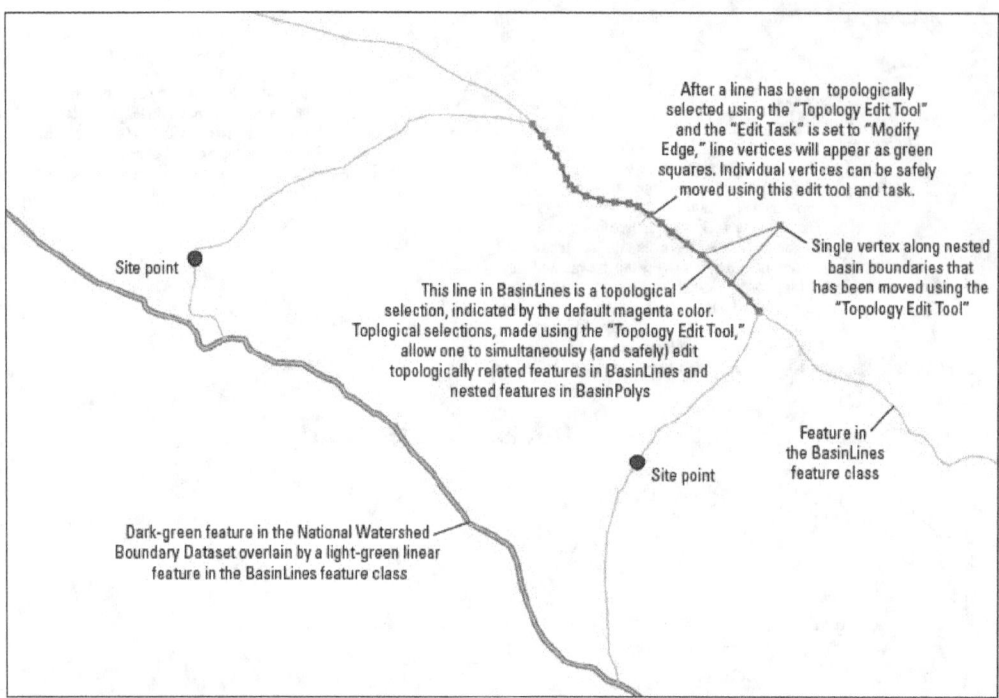

Base from U.S. Geological Survey digital data, 2012, 1:24,000
Albers equal-area conic projection
Standard parallels 37°30' N and 40°30' N
Central meridian 105°00' W

Figure 36.   ArcMap view of moving a single vertex using the "Topology Edit Tool" and the "Modify Edge" "Edit Task."

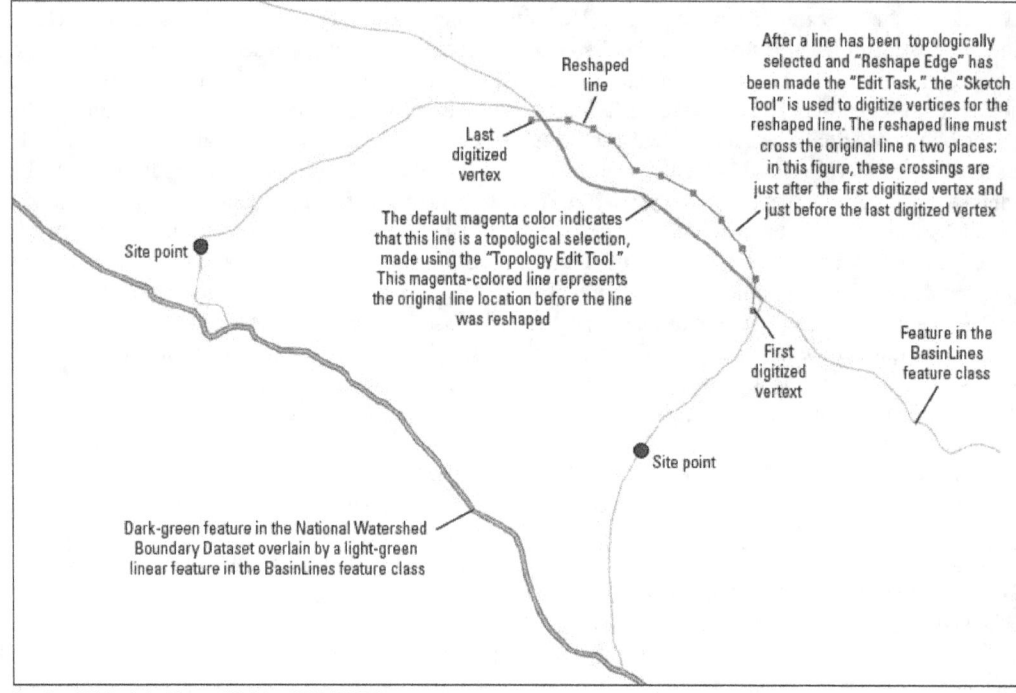

Base from U.S. Geological Survey digital data, 2012, 1:24,000
Albers equal-area conic projection
Standard parallels 37°30' N and 40°30' N
Central meridian 105°00' W

Figure 37.   ArcMap view of reshaping a line using the "Topology Edit Tool" and the "Reshape Edge" "Edit Task."

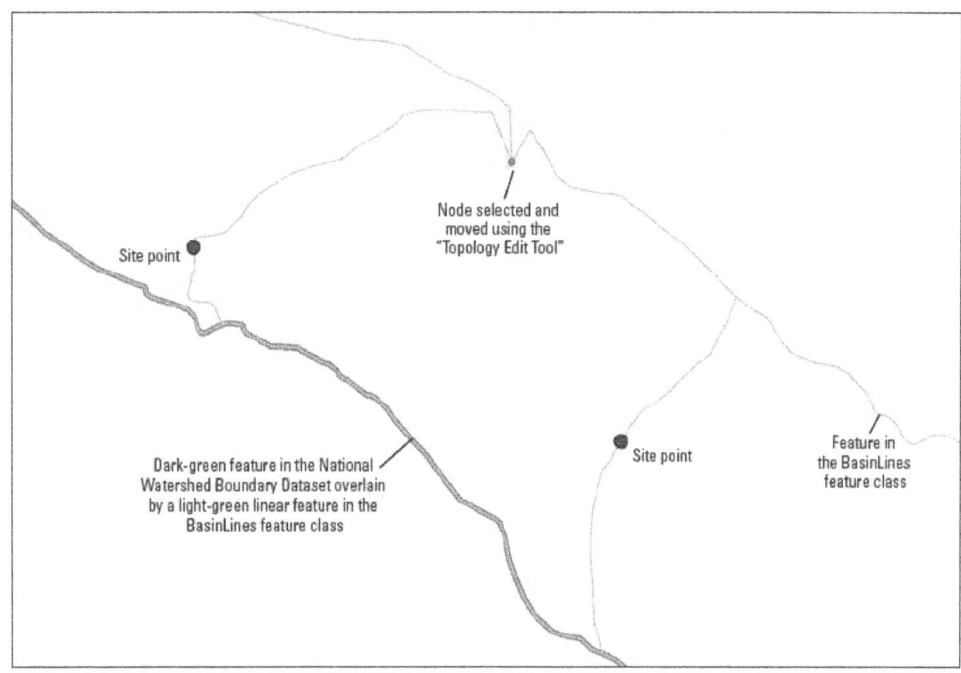

Base from U.S. Geological Survey digital data, 2012, 1:24,000
Albers equal-area conic projection
Standard parallels 37° 30′ N and 40° 30′ N
Central meridian 105° 00′ W

Figure 38.   ArcMap view of Moving a node using the "Topology Edit Tool" and the "n" key on the keyboard.

should be no errors or potentially only a few "BasinLines Must Not Intersect Or Touch Interior Point" errors, which can be repaired by right clicking on each error and clicking on the option to "Split."

Moving the End of a Line to a New Location

Sometimes the end of a line needs to be moved to a completely different location using a workflow called "splitting nodes." This situation can arise after basin review when it is determined that a digitized line or a WBD line needs to be corrected (amended). If there is a WBD line that needs to be amended, a new domain should be added to the domain values in ArcCatalog to flag the line for submission to the appropriate WBD steward for the state.[25]

1. To see where the nodes are located, click "Editor" on the Editor toolbar, select "Options…" from the pull-down menu, select the "Topology" tab, and at the bottom of the "Editing Options" dialog window check the checkbox to show "Unselected Nodes," and select "OK." If the nodes do not display, right click once anywhere in the map view using the "Topology Edit Tool" (fig. 19).

2. Splitting nodes is easier if the snapping tolerance is set to a relatively large value. To increase the snapping tolerance, click on "Editor" on the Editor toolbar, select "Options…" in the pull-down menu, and, in the textbox in the "Editing Options" window for "Snapping tolerance," type in 25 or 30 map units.

3. The goal of this exercise is to move a line and its endpoint to a location that more correctly follows the topographic drainage divide. If a node does not yet exist where the line endpoint needs to be relocated, a new node point must be created where the endpoint will be moved. The new node is created by splitting a line feature. To make a new "destination" node, make BasinLines the only selectable layer, set to snap to the BasinLines "Vertex," use the "Edit Tool" (fig. 14) to select an arc (using a regular type of feature selection), and use the "Split Tool" to split the arc and to make a new node. After splitting the arc, select the "Topology Edit Tool" on the Topology toolbar (fig. 19) and click once on the screen if the new node location is not being displayed (fig. 39).

4. Set the snap feature by selecting "Editor" on the Editor toolbar, and select "Snapping" from the pull-down menu to view the "Snapping Environment" window. Check the box in this window to allow lines to snap only to the "End" of the BasinLines feature class.

5. To relocate the node and the end of the line, zoom out so that the present line location and the new or "destination "node location are both visible. Then, select the "Topology Edit Tool," and simultaneously press "n" on the keyboard and draw a little selection box around

[25]Refer to *http //www.ncgc.nrcs.usda.gov/products/datasets/watershed/ coordinators.html* for the names of WBD State Stewards.

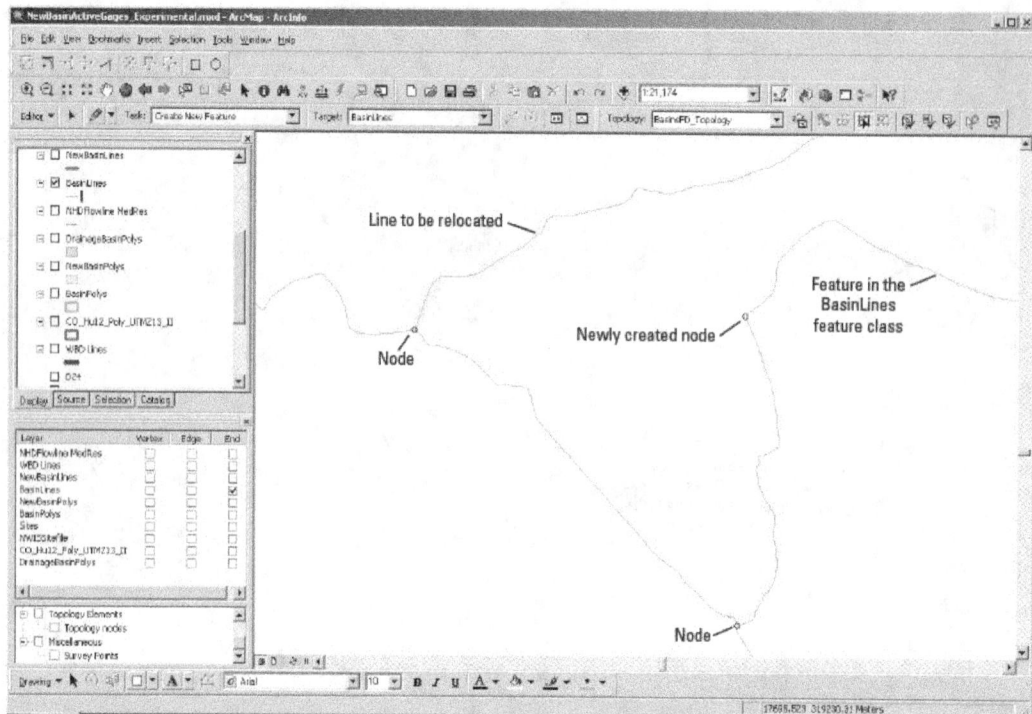

Base from U.S. Geological Survey digital data, 2012, 1:24,000
Albers equal-area conic projection
Standard parallels 37° 30' N and 40° 30' N
Central meridian 105° 00' W

Figure 39.   ArcMap view of a newly created node.

the node at the end of the line to be moved. The node should appear as a small magenta-colored dot, and no lines should yet be selected. If both node and an adjacent line are selected, click away from the area and repeat the node selection process using the "n" key until only a magenta node point is selected. Then, select the line (adjoining the node) that needs to be moved by pressing the "e" keyboard key (for 'edge') and the shift key (fig. 40).

6. With the "Topology Edit Tool" still active, press the "s" keyboard key and click again on the node to be moved. After a large cross-like cursor appears, release the "s" key and drag the node to the new or "destination" node location. If the snapping is set appropriately, the node and the attached line will move (fig. 41). Click away from the "destination" node location to unselect the topologically selected line and node.

7. If an error message appears and the node cannot be split, either the snapping is not set to an "End" of BasinLines, the snapping radius needs to be set a little higher, or the "Topology Edit Tool" is not active while the "s" keyboard key is selected and the feature is dragged to its new location.

8. After the node and line have been moved, select the basin boundary using the "Topology Edit Tool" and reshape the boundary using the "Reshape Edge" "Edit Task" (fig. 42).

9. Validate the entire topology, and there should be no errors. If validation of the topology takes longer than 10 or 15 minutes, refer to the "General Topological Editing Guidelines" section, and validate the topologically dirty area in sections. There should be no errors or possibly only a few "BasinLines Must Not Intersect Or Touch Interior Point" errors, which can be repaired by right clicking on each error and clicking "Split."

Cutting Polygons

If a basin polygon has been delineated incorrectly, it can be cut to a desired size. This sort of editing must be performed carefully if it affects multiple adjacent or nested basins. Because multiple polygons can be affected (or damaged) by this type of editing, it is wise to make a copy of the geodatabase before trying to cut boundaries of polygons already in the archive.

1. First, segregate a single basin to be cut using a feature definition query. Right click on the BasinPolys feature class, select "Properties…," and select the "Definition Query" tab view. Click on the "Query Builder" button. In the "Query Builder" window, create an SQL "where clause" to

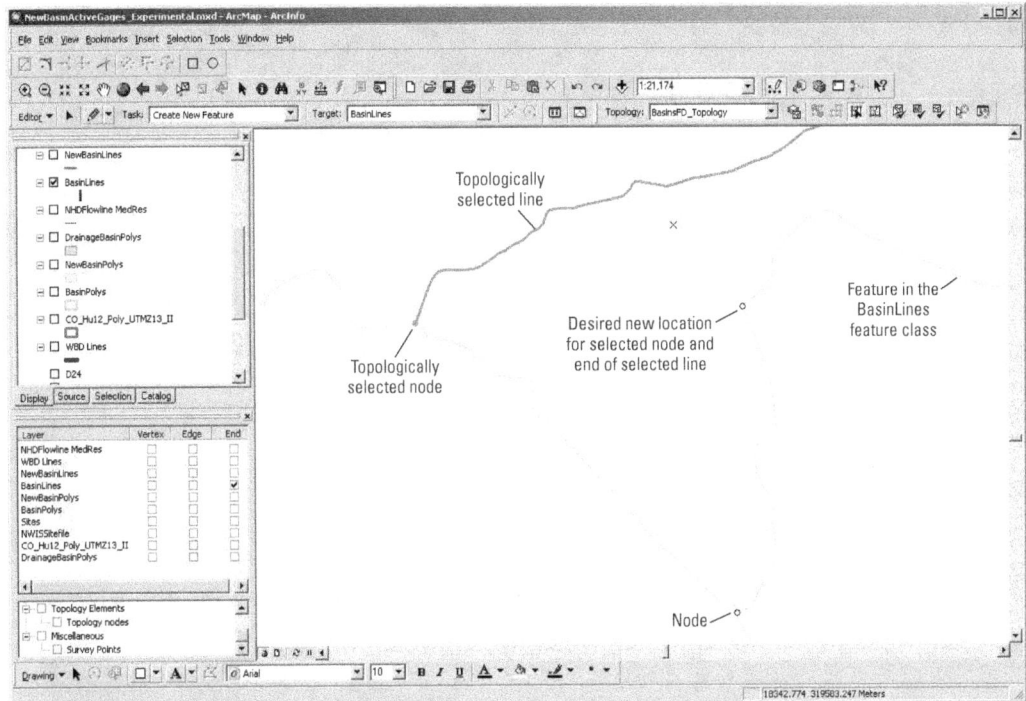

Base from U.S. Geological Survey digital data, 2012, 1:24,000
Albers equal-area conic projection
Standard parallels 37° 30' N and 40° 30' N
Central meridian 105° 00' W

Figure 40.   ArcMap view showing two topological selections: a line and a node endpoint.

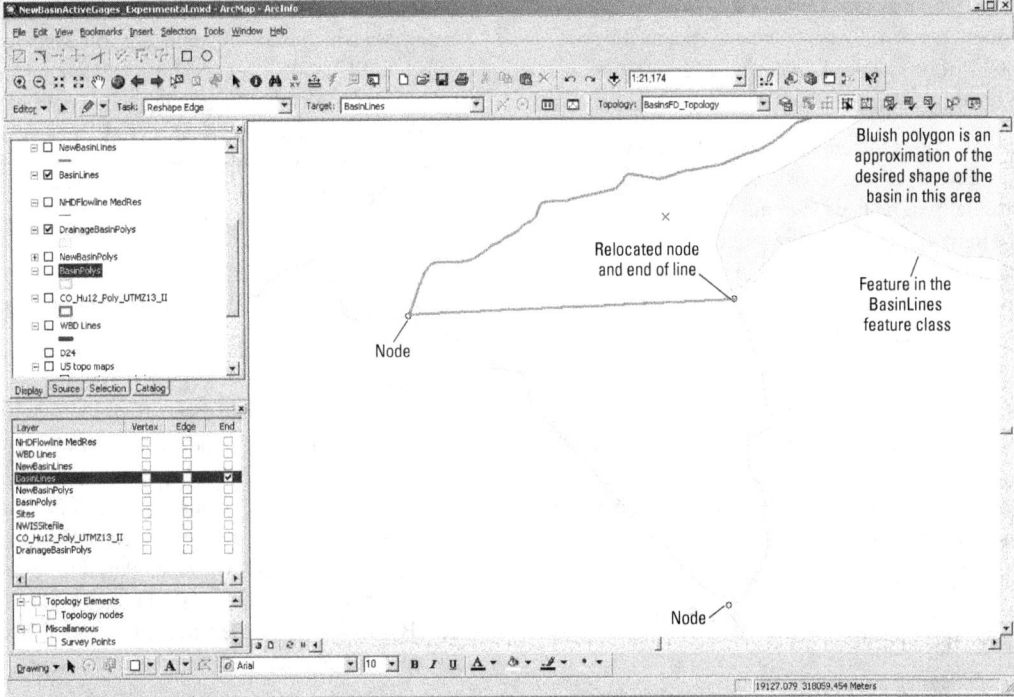

Base from U.S. Geological Survey digital data, 2012, 1:24,000
Albers equal-area conic projection
Standard parallels 37° 30' N and 40° 30' N
Central meridian 105° 00' W

Figure 41.   ArcMap view showing the selected line and node after they both have been moved.

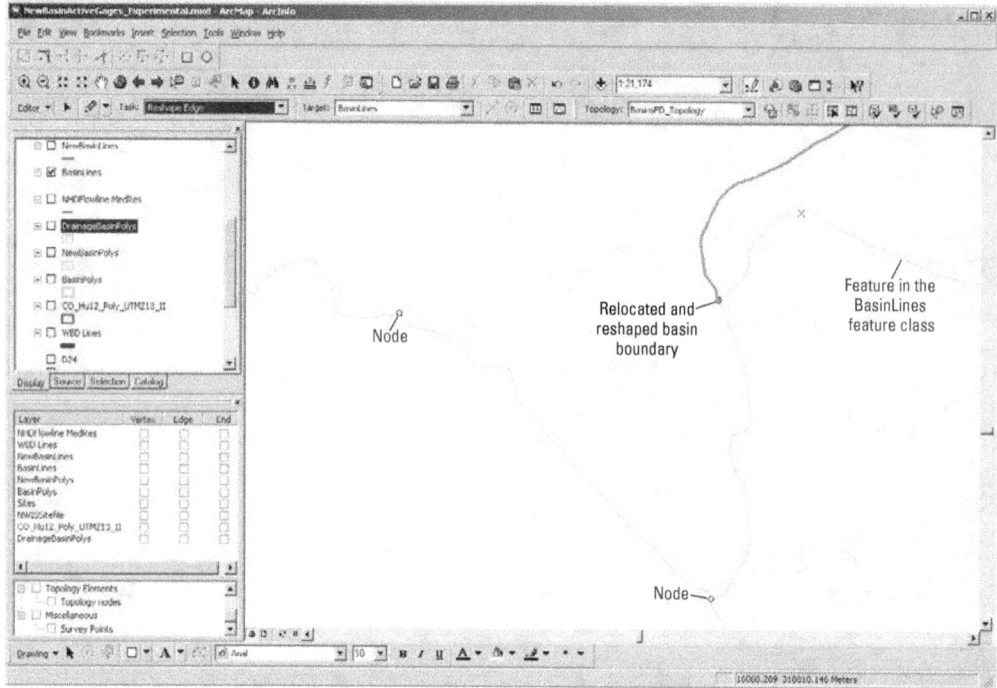

Base from U.S. Geological Survey digital data, 2012, 1:24,000
Albers equal-area conic projection
Standard parallels 37° 30' N and 40° 30' N
Central meridian 105° 00' W

Figure 42.   ArcMap view of showing final reshaped, relocated basin boundary.

select a single basin by "SiteName" field, and click "OK," and then "OK" to close the "Definition Query" window. After the query is run, only the one basin specified by the "SiteName" will be visible. An example of the feature definition query is shown in figure 43.

2. Make the WBD lines the only selectable layer and set the edit "Target Layer" to NewBasinLines. Select all WBD lines that will be needed to cut the polygon to the revised numeric drainage area and copy them into NewBasinLines. After visually making sure that there are no gaps in the cutter line, make NewBasinLines the only selectable layer, select all the features, and merge them into a single cutter line (fig. 44).

3. Make BasinPolys the only selectable layer, and select the single feature (selected by the feature definition query) in BasinPolys. Then, right click on the NewBasinLines cutter line, select "Replace Sketch," and then right click once on either side of the cutter line and select "Finish Sketch."

4. Right click on BasinPolys, select "Properties…" and delete the SQL "where clause" in the "Definition Query" window, and click "OK."

5. If the cut polygon outside the true basin boundary can simply be deleted (if it does not belong to an adjacent basin polygon), select and delete this extraneous polygon (it will be the last feature in the BasinPolys attribute

table). If the extraneous polygon should be merged with an adjacent basin, select the polygon and the adjacent basin into which the polygon needs to be merged, and select "Edit," and "Merge," and select the named adjacent basin in the merge window as the target of the merge operation. If several basins adjacent to the basin that was cut need to have this 'extraneous' polygon added, first make multiple copies of the extraneous polygon in BasinPolys and merge each copy sequentially with each adjacent basin.

6. Select and delete all features in NewBasinLines.

7. Validate the entire topology or validate to remove dirty areas (refer to the "General Topological Editing Guidelines" section), and there should be two types of errors: a "BasinPolys Boundary Must Be Covered By BasinLines Polyline" error and one or more "BasinLines Must Be Covered By Boundary Of BasinPolys Polyline" errors. Right click on each of the "BasinPolys Boundary Must Be Covered By BasinLines Polyline" errors and select "Create Feature." Right click on each of the "BasinLines Must Be Covered By Boundary Of BasinPolys Polyline" errors and select "Subtract."

8. Validate topology, and there may be a few "BasinLines Must Not Intersect Or Touch Interior Point" errors. Right click on these and select "Split." Validate, and there should be no more errors.

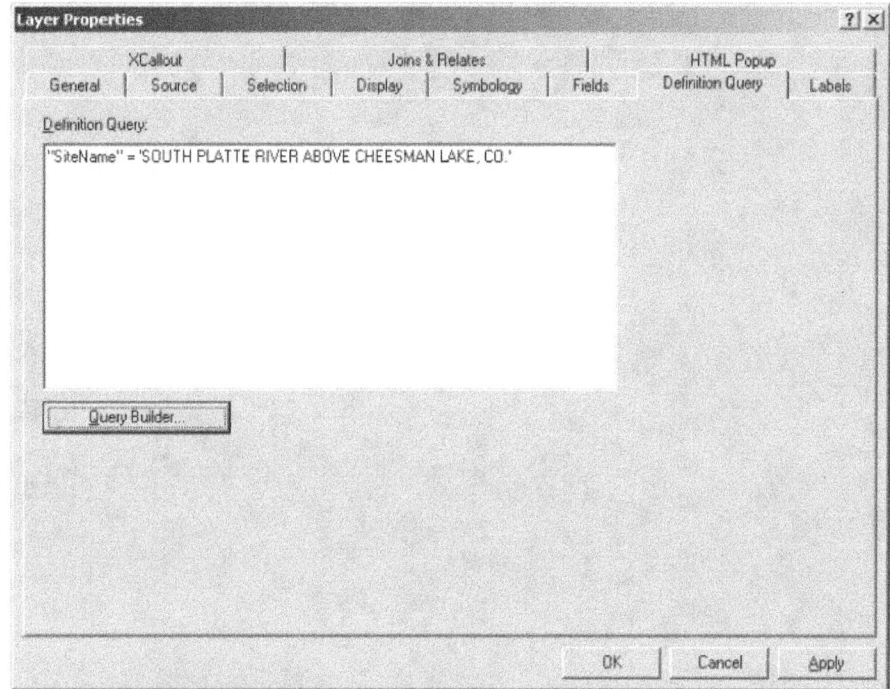

Figure 43.   Definition Query window showing one selected drainage basin.

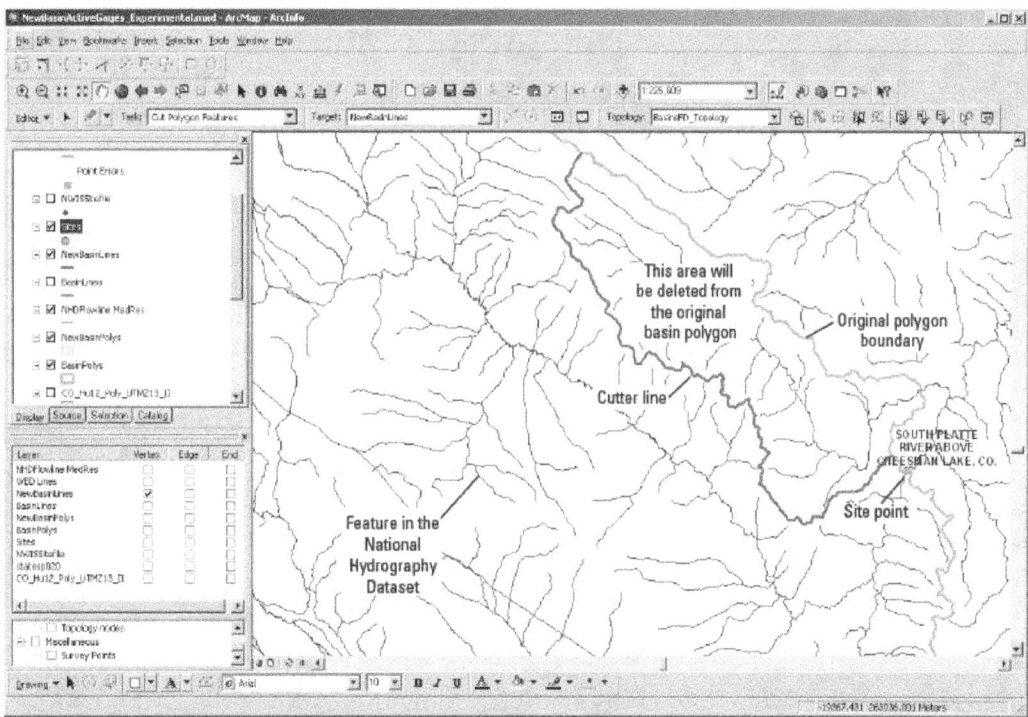

Base from U.S. Geological Survey digital data, 2012, 1:24,000
Albers equal-area conic projection
Standard parallels 37° 30' N and 40° 30' N
Central meridian 105° 00' W

Figure 44.   ArcMap view of a single basin polygon and cutter line.

Appendix 2.   Guidelines for Geodatabase Scope, Drainage-Basin Delineation, and Geodatabase Management

Introduction

This appendix contains guidelines used by the U.S. Geological Survey (USGS) Colorado Water Science Center (CWSC) regarding geodatabase scope, drainage-basin delineation, and geodatabase management. Development of these CWSC guidelines relied on several published reference standards for delineation of hydrologic unit boundaries (Federal Interagency River Basin Committee, Subcommittee on Hydrology, 1951; Office of Water Data Coordination, 1978; U.S. Geological Survey and U.S. Department of Agriculture, Natural Resources Conservation Service, 2009).

Guidelines for Geodatabase Scope

Geographic Scope

Basins generally are stored only for the state that is the purview of the water science center or for basins inside the hydrologic subregions that at least partially occupy the state. Exceptions are made for regional studies that include the state, such as USGS National Water Quality Assessment Program (NAWQA) regional studies. Polygonal basin areas also are not terminated at political boundaries but should represent the full, hydrologic drainage basin.

Types of Basins Stored

As defined by the Federal Interagency River Basin Committee, Subcommittee on Hydrology (1951), "The drainage area of a stream at a specified location ordinarily may be defined as that area, measured in a horizontal plane, which is enclosed by a topographic divide such that direct surface runoff from precipitation normally would rain by gravity into the river basin above that point." According to the delineation standards developed by the USGS and the Natural Resources Conservation Service (U.S. Geological Survey and U.S. Department of Agriculture, Natural Resources Conservation Service, 2009), basins are delineated based on only topography and hydrologic features. In other words, basins delineated for a site point are always topographic (or geomorphic) basins and are not delineated with consideration of sources of water arriving at the site point.

As a corollary to the above rule, drainage basins are generally not delineated for artificial water features (such as ditches, laterals, or canals). However, basins can be delineated for channelized streams or for ditches that were once stream channels. Basins also are not delineated by incorporating information about groundwater movement, and are not delineated for springs.

The polygonal basin areas stored are used to generate total numeric drainage areas, but polygonal basin areas can also be delineated for contributing and noncontributing areas.

Noncontributing polygonal basin areas consist of areas of internal drainage and may include pothole areas in glaciated plains, closed basins, playas, cirques, depression lakes, dry lakebeds, karst features, sand dunes, or similar landforms. Larger noncontributing polygonal basin areas can, in part, be derived from the National Watershed Boundary Dataset (WBD) by using a Structured Query Language (SQL) statement "where clause" like "HU_12_TYPE = 'C'" or "HU_12_DS = 'CLOSED BASIN'."

Basin Delineation Guidelines

Basin Site-Point Location

1. The basin-site point represents a streamgage-orifice location, the location of a water-quality sampling site, a stream confluence, or some other USGS study- or project-related site location. Before a drainage basin is delineated, the site location should be determined accurately using a sitefile built from National Water Information System (NWIS) coordinates, the NWIS coordinates themselves, the NWISWeb location map (U.S. Geological Survey, 2012a), the site description from the Annual Water Data Report (U.S. Geological Survey, 2012d), or from information supplied by USGS Water Science Center personnel who have recently visited the site location.

2. A secondary basin-site point may also be stored to represent the true topographic outlet for the basin. The geodatabase accompanying this report does not include this second, optional site-point GIS dataset. This secondary site point should be located on the high-resolution National Hydrography Dataset (NHD) flowline closest to the actual streamgage location. A GIS dataset of USGS streamgages snapped to the high-resolution NHD is available from the NHD Web site (U.S.Geological Survey, 2012b). If the drainage basin is being determined for a confluence, the site point should be snapped to a high-resolution (1:24,000-scale) NHD flowline above a stream junction shown on the NHD.

3. If a stream channel or braid has moved and is not in the same location shown on the digital historical topographic map, the basin-site point may be positioned at a global-positioning-system- (GPS-) measured location (whether the channel or braid is shown on the digital topographic map or not). The Environmental Sciences Research Institute (ESRI) "Imagery" map service (Environmental Research Systems Institute, 2012b) can be useful in positioning such a revised streamgage location because the gage house may be visible in the satellite imagery.

4. The basin boundary should be digitized from and snapped to the basin-site point.

Source GIS Datasets Used for Basin Delineation

Two source GIS datasets are generally used for delineating new drainage basins: (a) lines in the WBD or (b) topographic contours from 1:24,000-scale digital raster graphics (DRG) maps.

1. If the basin-perimeter delineation method is being used, the first lines copied into a new basin perimeter should be any lines already archived in BasinLines. After integrating any archived lines, lines in the WBD should be copied as the new basin perimeter is being built. If the cut-polygon method is used, the WBD is the polygon source.

2. Digital topographic maps at a scale of 1:24,000 are the ultimate reference for line work, because the drainage basins delineated are topographic basins.

3. Once the new polygonal basin area has been added to the archive, polygonal basin areas in the archive should be edited only if requested by the Water Science Center Data Section, which then should be notified of resulting changes in the numeric drainage area after the polygonal basin area has been edited.

Digitizing New Lines

New or added lines are added from the basin-site point up to the nearest upstream WBD line. If the basin is small and lies within a single 12-digit WBD hydrologic unit, the entire basin may have to be digitized.

1. Digitizing of new lines should start at the basin-site point. Starting at the basin-site point generally helps when making decisions in areas where topographic interpretations are difficult. Drainage divides should be traced from the basin-site point on the digital topographic maps along divides indicated by contour lines.

2. As often as possible, digitized lines should be perpendicular to topographic contours.

3. Basin boundary lines cannot cross other basin boundaries. Polygonal basin areas can nest and their basin boundaries can reuse existing lines, but the boundaries of adjacent basins cannot cross.

4. Accuracy is keyed to the 1:24,000 DRG topographic maps. Because adjacent and nested basins may use the same digitized basin boundary, basin boundaries should be located as accurately as possible. Newly digitized lines should be traced from 1:24,000-scale DRG maps at a tight zoom scale to correctly position the topographic drainage divide. In mountainous areas or where the location of the drainage divide separating adjacent drainage basins is clear, the line should vary no more than 1/8th inch from an obvious ridgeline at a scale of 1:10,000. In areas of low relief, positioning of the drainage divide is more interpretive, and it is possible to have multiple interpretations. Generally, digitizing should be done at a map scale set to at least 1:10,000 and probably to between 1:6,000 and 1:10,000. In areas of low relief, however, it is useful to zoom out to a smaller scale to visualize the drainage from the topography. In such areas, shaded-relief topographic maps, such as the ESRI "Shaded Relief" map service, can be helpful.

Topographic Interpretation

Guidelines for Anthropogenic Features

Drainage basins delineated for the geodatabase are topographic basins. Because the delineator follows the topographic contours to delineate a drainage basin, anthropogenic features can be ignored unless they are accompanied by local changes to topographic contours. Guidelines have been developed for the following specific anthropogenic features.

1. For in-stream reservoirs, the delineator determines the lowest spill point, and the entire reservoir drainage basin is assigned to the spill-point drainage basin. The basin boundary is drawn around reservoirs, never through reservoirs. For non-instream reservoirs, the basin boundary should be traced around the feature; never through it. For reservoirs near a drainage divide, assign the entire reservoir to the basin into which it would drain if it overfilled.

2. Unless there are topographic contours surrounding diversion features (such as ditches, canals, or outfalls), these artificial drainage features are ignored.

3. Freeways, roads, and railroads are considered during basin delineation only if they form a topographic barrier to water movement. If road features are present and contours indicate the presence of berms around such features, the delineator should strictly follow the contours. If there are no contour changes near such features, they are ignored, and the basin boundary may cross them at a topographically appropriate point.

4. Unless there is local drainage pattern information or changes in contours, buildings and other manmade structures are ignored.

5. Delineation in urban areas can sometimes be facilitated by referring to older versions of topographic maps that show the topographic contours before urbanization occurred.

6. Interpolation between contours may be indicated by reference to old trails, roads, and firebreaks in forested areas, all of which often follow drainage divides (Federal Interagency River Basin Committee, Subcommittee on Hydrology, 1951). Surveying points indicated on the topographic maps are most commonly located on ridge crests or peaks to maximize line-of-sight for surveying and are often at or close to a drainage divide.

Guidelines for Natural Features

Natural features are treated using the following rules when digitizing new lines from the site point to the WBD lines. Rules for topographic interpretation in areas such as coastlines and karst areas are discussed in several references (Federal

Interagency River Basin Committee, Subcommittee on Hydrology, 1951; U.S. Geological Survey and U.S. Department of Agriculture, Natural Resources Conservation Service, 2009).

1. For broad, alluvial valleys where groundwater inflow is parallel to the river, the basin boundary should be drawn perpendicular to the topographic contours even if a map of the groundwater flow path exists.

2. Basin boundaries should not cross ponds or lakes. Such features should be assigned to the side of the basin that they would drain to if they were filled.

3. Drainage basins generated in areas of low relief can be sometimes more easily delineated in such areas if the delineator first delineates a "guide basin" derived from 10-m National Elevation Dataset (NED) (U.S. Geological Survey, 2012f) or from StreamStats (U.S. Geological Survey, 2012e). Shaded relief maps (such as the ESRI "Shaded Relief" map service) (Environmental Research Systems Institute, 2012b) are also useful for picturing the direction of runoff in the basin. Because such NED-derived basins can be inaccurate in areas of low relief, such basins are used to help envision the general basin location with reference to topography. Zooming out and using smaller scale DRG maps can help to perceive gross trends in topography.

4. Hummocky topography, like that present in areas of sand dunes or karst terrain, is extremely difficult to interpret. Again, use of smaller scale DRG maps that combine topographic contours with shaded relief can help to visualize the overall topographic highs and lows.

5. For braided streams, the basin boundary should be developed for the braid nearest the streamgage or water-quality sampling site. The braid on which the streamgage or site is located may be treated as the main channel, and other braids should be ignored when the basin boundary is delineated. If necessary, subtle local topography near other braids may also be ignored.

6. When the headwaters of two or more basins is in the same depression (or swamp) along a ridgeline, each respective stream is assumed to drain that portion of the depression lying closer to it than to any other stream, unless more definite information is available from field inspection.

7. If topography has changed near a site location since the publication of the USGS 1:24,000-scale topographic map, the basin boundary can be developed using the new topographic contours or using the new location of the stream channel as shown on satellite imagery. This revised channel can be used as the main channel if the stream is braided.

Geodatabase Management Guidelines

Basin boundaries are developed for the geodatabase in two stages: delineation and finalization. Each stage can entail editing, which requires "checking out" the geodatabase, which is described here. To prevent situations where there are multiple edited versions of the geodatabase that have to be reconciled, the geodatabase has a chain of custody and there is a protocol for any use that could lead to editing.

1. The most current or master version of the geodatabase resides in a file storage area that is backed up at least weekly. Other backup versions may also be stored in other locations.

2. The geodatabase should be public and available to everyone. The number of people who edit the geodatabase, however, should be limited to as few as possible to ensure high data quality.

3. If a local copy of the geodatabase is copied to a personal computer, the working version geodatabase should be renamed to indicate that it is "checked out" to signal other potential users that the geodatabase is being edited (for example, rename the "COBasins.gdb to "COBasins_ checkedout.gdb"). If the working version is edited during an ArcMap session, the renaming protocol is not required because the geodatabase will be locked to other editing. Before copying the edited version back to the main storage area, either validate the entire topology or sequentially validate topologically dirty areas until they are eliminated and ensure that no topological errors are present.

www.ingramcontent.com/pod-product-compliance
Lightning Source LLC
Chambersburg PA
CBHW081607170526
45166CB00009B/2860